South East Asia has for many centuries occupied a pivotal position in the wider Asian economy, linking China and the Far East with India and the Middle East, and since the early 1500s the region has also played a major role in the world-economy. *South East Asia in the world-economy* is the first textbook survey of the area's interaction with these wider regional and international structures.

Professor Chris Dixon demonstrates how South East Asia's role has undergone frequent and profound change as a result of the successive emergence and dominance of mercantile, industrial and finance capital. He shows how the region has developed as a supplier of luxury products, such as spices; as a producer of bulk primary products; and how, since the mid-1960s, it has become a major recipient of investment and a favoured location for labour-intensive manufacturing operations, producing goods for European and American markets. The author examines how these phases in the evolution of the international economy have been reflected in the relations of production and in the spatial pattern of economic activity. He also discusses how the progressive integration of South East Asia in the world-economy has established the dominance of a small number of core areas and produced a pattern of uneven development throughout the region. In a concluding chapter, Chris Dixon explores the prospects for South East Asia in the 1990s in the light of the restructuring of the world-economy.

Geography of the World-Economy

Series editors:

A geography without knowledge of place is hardly a geography at all. And yet traditional regional geography, underpinned by discredited theories of environmental determinism, is in decline. This new series *Geography of the World-Economy* will reintegrate regional geography with modern theory and practice – by treating regions as dynamic components of an unfolding world-economy.

Geography of the World-Economy will be a textbook series. Individual titles will approach regions from a radical political–economic perspective. Regions have been created by individuals working through institutions as different parts of the world have been incorporated in the world-economy. The new geographies in this series will examine the ever-changing dialectic between local interests and conflict and the wider mechanisms, economic and social, which shape the world system. They will attempt to capture a world of interlocking places, a mosaic of regions continually being made and remade.

The readership for this important new series will be wide. The radical new geographies it provides will prove essential reading for second-year or junior/senior students on courses in regional geography, and area and development studies. They will provide valuable case-studies to complement theory teaching.

South East Asia in the world-economy

Also published in this series

The United States in the world-economy

John Agnew

The United States in the World-Economy is a major new textbook survey of the rise of the United States within the world-economy, and the causes of its relative decline. With the USA being the dominant state in the contemporary world-economy, it is vital to understand how it got where it is today, and how it is responding to the current global economic crisis. Professor Agnew emphasizes the divergent experiences of different regions within the USA, and in so doing provides a significant 'new' regional geography, tracing the historical evolution of the USA within the world-economy, and assessing the contemporary impact of the world-economy upon and within it. No existing treatment covers the subject with equivalent breadth and theoretical acuity, and the guiding politico-economic framework provides a coherent radical perspective within which the author undertakes specific regional and historical analysis. *The United States in the World-Economy* will prove required reading for numerous courses in regional geography, area studies and the geography of the United States.

South East Asia in the world-economy

Chris Dixon

Professor of Geography, City of London Polytechnic

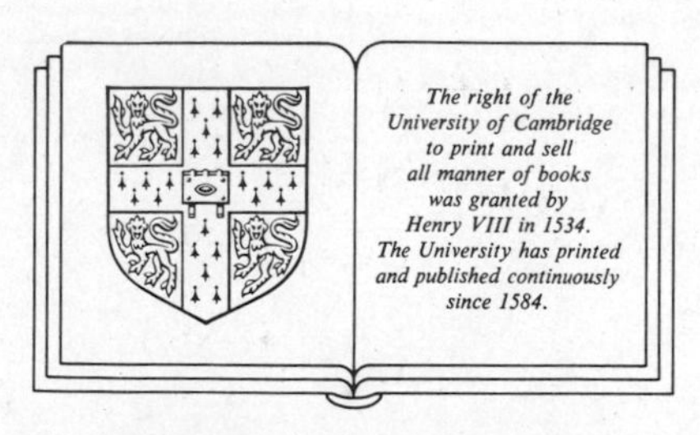

CAMBRIDGE UNIVERSITY PRESS

Cambridge

New York Port Chester

Melbourne Sydney

Published by the Press Syndicate of the University of Cambridge
The Pitt Building, Trumpington Street, Cambridge CB2 1RP
40 West 20th Street, New York, NY 10011, USA
10 Stamford Road, Oakleigh, Melbourne 3166 Australia

First published 1991

Printed in Great Britain at the University Press, Cambridge

British Library cataloguing in publication data

Dixon, Chris
South East Asia in the world-economy – (Geography of the world-economy).
1. South-east-Asia. Economic development
I. Title II. Series
330.959

Library of Congress cataloguing in publication data

Dixon, C. J. (Chris J.)
South East Asia in the world-economy / Chris Dixon.
p. cm. – (Geography of the world-economy)
Includes bibliographical references and index.
ISBN 0 521 32035 6 (hardback) – ISBN 0 521 31237 X (paperback)
1. Asia, Southeastern – Economic conditions. 2. Asia, Southeastern – Commerce. I. Title. II. Series.
HC441.D59 1991
330.959 – dc20 90–35861
CIP

ISBN 0 521 32035 6 hardback
ISBN 0 521 31237 X paperback

CE

Contents

List of figures *page* vii
List of maps ix
List of tables x
Acknowledgements xiii
Sources and place names xiii
List of abbreviations xiv

1 Contemporary South East Asia in the world-economy 1

2 Pre-colonial South East Asia 35

3 Western penetration: from trade to colonial annexation 57

4 Uneven development: the establishment of capitalist production 85

5 Development strategies and the international economy 149

6 South East Asia in the late twentieth century: problems and perspectives 217

Notes 227
References 247
Index 274

Figures

1.1	Annual net inflows of foreign direct investment, 1960–86	*page* 14
1.2	Commodity prices, 1948–87	24
1.3	Thailand, terms of trade, 1961–86	30
1.4	Balance of payments on current account as a percentage of GDP, 1965–86	31
2.1	Seasonal activity in Thai rice cultivation	38
4.1	Growth of coffee and sugar production in the Netherlands East Indies, 1807–1986	86
4.2	Growth of South East Asian primary exports, 1840–1940	87
4.3	Expansion of rice cultivation, 1850–1940	91
4.4	Thai rice yields, 1917–50	94
4.5	Capital investment in South East Asia during the late 1930s	113
4.6(a)	Origins of South East Asian imports, share of total value, 1938	116
4.6(b)	Destination of South East Asian exports, share of total value, 1938	117
5.1	Exports of the Asian Newly Industrialising Countries	164
5.2	Composition of Singapore's exports, 1975–85	165
5.3	Percentage changes in industrial production of Singapore and the other East Asian Newly Industrialising Countries	169

Maps

1.1	South East Asia	*page* 2
2.1	Structure of South East Asia	37
2.2	Major early historical settlement centres and congenial habitats	39
2.3	Languages of indigenous peoples	43
2.4	South East Asia: political divisions in the late fifteenth century	46
3.1	South East Asian products during the late sixteenth century	58
3.2	Iberian possessions and sailing routes during the sixteenth century	60
3.3	VOC possessions and sailing routes during the seventeenth and eighteenth centuries	65
3.4	Expansion of Dutch territorial control	67
3.5	Expansion of European control	70
3.6	The establishment of British control in the Malay Peninsula	74
4.1	Economic development in the Malay Peninsula, 1941	125
5.1	Indonesian special development areas	196

Tables

1.1	South East Asian economies in perspective, 1986	*page* 5
1.2	The attraction of South East Asia for labour-intensive manufacturing investment	6
1.3	Growth of GDP, average annual percentage rate	7
1.4(a)	South East Asian primary production, 1986	7
1.4(b)	National percentage contributions to South East Asian primary production and export, 1986	8
1.5	Defence expenditure, 1986	11
1.6	Percentage share of net ASEAN foreign investment	13
1.7	Sources of South East Asian foreign direct investment, percentage share	16
1.8	Estimated sectorial distribution of foreign direct investment, percentage share	17
1.9	Growth of merchandise trade, 1960–86 (percentage annual average rate of growth)	19
1.10	Integration into the world economy: the value of foreign trade as a percentage of GNP	20
1.11(a)	Export intensities of South East Asian countries with Japan, USA, European Community and ASEAN, 1984	20
1.11(b)	Import intensities of South East Asian countries with Japan, USA, European Community and ASEAN, 1984	21
1.12	Changing composition of the value of international trade, 1965–86	22
1.13	Intra-ASEAN exports as a percentage of the total value of exports, 1970–87	23

1.14	Changing structure of the South East Asian economies	25
1.15	Urbanisation	27
1.16	Average annual growth rates of population by World Bank income group (percentage)	28
1.17	Population distribution in Indonesia, 1983	29
1.18	Growth of foreign debt, 1970–86	32
1.19	Annual rate of growth of GDP, 1980–7	33
2.1	Estimates of ethnic composition	44
4.1	Thai exports, *c.* 1850	88
4.2	Thai exports, 1867–1950	89
4.3	Occupancy of land in Lower Burma, 1925–6 to 1937–8	100
4.4(a)	Netherland East Indies, principal plantation crops in 1930	102
4.4(b)	Netherland East Indies, the contribution of estates to total production and export earnings in 1930	102
4.5	South East Asian primary production, 1937–40 in metric tons	110
4.6	Distribution of Chinese and Chinese investment in South East Asia during the late 1930s	111
4.7	Distribution of French investment on the basis of new capital issues of Indochinese corporations, 1924–30 and 1931–8	118
4.8	Ownership of the manufacturing sector in the Netherlands East Indies during the late 1930s	120
4.9	Value of exports, Java versus the Outer Islands, 1870–1940	124
5.1	Tariff protection in South East Asia, 1978	154
5.2	Export incentives	155
5.3	Singapore, employment in manufacturing and processing industries, 1955	158
5.4	Commercial banks, merchant banks and Asian Currency Units in Singapore, 1968–84	166
5.5	Government expenditure as a percentage of GDP, 1965–87	171
5.6(a)	Foreign ownership of the Thai financial sector, 1979	178
5.6(b)	Ownership of various sectors of Thai industry, 1978–9	178
5.7	Thailand, changing composition of non-energy imports, percentage of total values	181

5.8(a)	Peninsular Malaysia: employment by occupation and race, 1970	182
5.8(b)	Peninsular Malaysia: mean household incomes, 1970	182
5.8(c)	Peninsular Malaysia: urban residence by ethnic group, 1970	182
5.9	Malaysia development expenditure 1956–90, percentage distribution	184
5.10	Indonesian development budget, 1974–89, percentage share of sectors	197
5.11(a)	Indonesia, sponsored transmigration programme, 1950–84	199
5.11(b)	Indonesia, destination of sponsored migrants, 1950–84	199
5.12	Changing incidence of poverty in South East Asia, World Bank studies	210
5.13	Changes in the incidence of poverty and the level of income inequality in Thailand, 1962–81	210
5.14(a)	Thailand's regional per capita gross product as a percentage of the national per capita GDP (at constant 1962 prices)	212
5.14(b)	Thailand's Regional per capita gross product as a percentage of national per capita GDP (at constant 1972 prices)	213
5.15	Malaysia, changing incidence of poverty, 1976–85	214
6.1	Hourly compensation of production workers in the Asian electronics industry	221

Acknowledgements

The author wishes to thank David Drakakis-Smith and Peter Taylor for reading and commenting on an earlier draft, Sue Shewan for typing the manuscript and the staff of the City of London Polytechnic Cartography Unit for the production of the maps and diagrams.

A note on place-names

The period covered by this book has seen many changes in the names of countries and places. Contemporary and current names have been used as the sense of the discussion dictated. In some instances consistency and strict historical accuracy have been sacrificed to clarity. In particular, Thailand has been used throughout the book except in citations, although the official name was Siam until 1939, and again briefly at the end of the Second World War.

Abbreviations

ADB	Asian Development Bank
ASEAN	Association of South East Asian Nations
BOI	Board of Investment
CPF	Central Provident Fund
EC	European Community
EIC	East India Company
EOI	Export Orientated Industrialisation
EPZ	Export Processing Zone
ESCAP	Economic and Social Commission for Asia and the Pacific
FAO	Food and Agricultural Organisation
FELDA	Federal Land Development Agency
FTZ	Free Trade Zone
GDP	Gross Domestic Product
GNP	Gross National Product
IBRD	International Bank for Reconstruction and Development (World Bank)
IEAT	Industrial Estate Authority of Thailand
IMF	International Monetary Fund
ISI	Import Substituting Industrialisation
NEDB	National Economic Development Board
NEDCOL	National Economic Development Corporation Limited
NEI	Netherlands East Indies
NEP	New Economic Policy
NESDB	National Economic and Social Development Board
NIC	Newly Industrialising Country
PAP	Peoples' Action Party
UNDP	United Nations Development Programme

VOC Vereenigde Oostindische Compagnie (United East India Company, in English writing usually referred to as the 'Dutch East India Company')

1

Contemporary South East Asia in the world-economy

1.1 Introduction

South East Asia has played a major role in the world-economy. It has supplied key raw materials, provided markets for developed world goods, received investment and, most recently, multinational manufacturing. To a degree these roles have been sequential, reflecting the evolution of imperialism, the progressive incorporation of the region into the world-economy and the accompanying development of capitalist relations of production.

Western recognition of South East Asia (map 1.1) as a distinct entity is comparatively recent.[1] Until the 1940s a variety of terms were applied to these islands and peninsulas, for example, Further India, the Far Eastern Tropics and Indo-China. Thus, South East Asia was viewed as an eastward extension of India, a tropical appendage to the Far East or merely the area between India and China.[2] These terms strongly reflect the imperial structures that had been established by the early 1900s, as indeed the adoption of the term South East Asia during the 1940s reflected a radically different situation.

The emergence of South East Asia as a major sub-division of the world economy can be discerned during the 1930s. This was principally the result of the weakening of the established imperial structures. In terms of British policy, events in East Asia and moves towards some form of independence for India shifted the focus of activity towards Burma and the Malay Peninsula. Additionally, from the mid-1920s South East Asia's resources and markets had become a focus of increasing rivalry between the USA and Japan.

The new regional structure was consolidated by the Second World War and its immediate aftermath. In British official thinking the

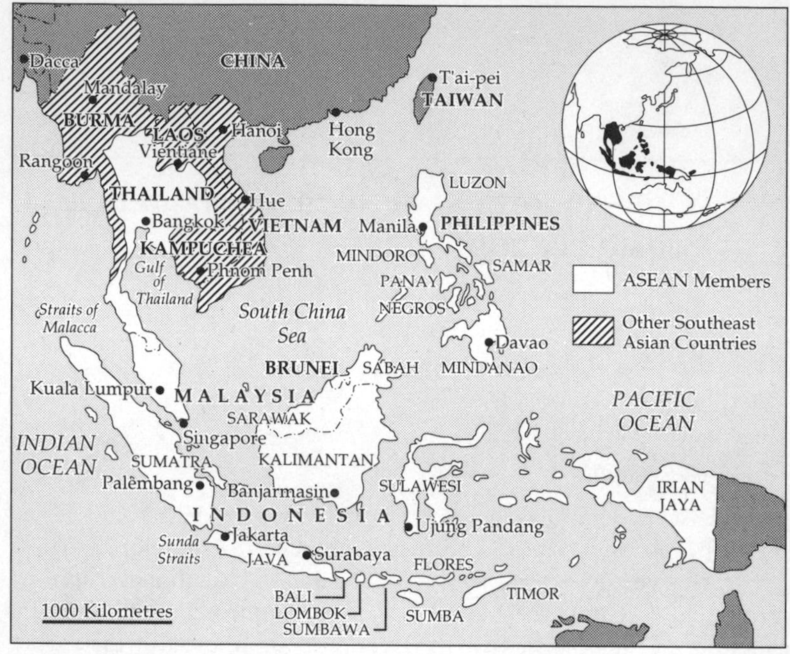

Map 1.1 South East Asia

concept of 'South East Asia' began to crystallise from 1945 (Smith, 1988). British activity 'East of Suez' came to focus on Singapore and Malaya. These territories, especially Singapore, became regarded in economic, strategic and diplomatic terms as the pivot of a 'South East Asian region'. More significantly, from 1947, South East Asian raw materials and markets begin to feature in American strategic thinking and policy statements. (McMahon, 1981:143). The subsequent emergence of American hegemony in the region is both a reflection and an integral part of the emergence of South East Asia as a major subdivision of the post-war world-economy.

Most writers, while accepting the comparatively recent recognition of South East Asia as a major region, have devoted considerable attention to its distinctiveness on cultural, linguistic and environmental grounds. Thus South East Asia is presented as comprising a complex human and physical mosaic which sets it apart from neighbouring lands. Attention is drawn to the region's physical configuration and location which made it from early times a focus of land and sea routes, gave it significance in the economic structures of pre-colonial Asia, and contributed to its distinctiveness. Thus for many writers South East Asia has long been one of the world's great 'zones of convergence'. Others have seen it as a 'shatter zone' of meeting and conflict (Spyekman, 1942). In similar vein the region has been described as the 'Balkans of the Orient' (Fisher, 1962).

Much of this writing is highly deterministic, implying that the strength of South East Asia's personality forced its belated recognition as a major world region. While not entirely denying the underlying significance of cultural and environmental factors in the interaction between South East Asia and the world-economy it is argued here that regional divisions are the product of international economic and political structures and processes external to the region. However, changes in recognised regional divisions do not automatically follow from changes in the structures that underpin them. Frequently regions and regional sub-divisions become institutionalised. Many of the widely accepted groupings are relics of imperialism's earlier phases and their continued use frequently places our analysis in a strait-jacket. In this South East Asia is no exception.

In terms of the world-economy South East Asia has since the 1940s been a far from unified region. Any analysis is dominated by the 'Vietnam' war and the divisions between the pro- and anti-Western states that have emerged. While for much of the period this division was a far from clear or fixed one, as Värynen (1988) has stressed it became polarised between Vietnam and the pro-capitalist group, and institutionalised with the establishment of ASEAN (Association of South East Asian Nations) in 1967.

The capitalist-socialist division in South East Asia is complicated by Burma which has in recent years generally adopted a non-aligned, isolationist position. To some extent Burma has attempted to keep on distant but friendly terms with both groups as witnessed by the hosting of Vietnamese–ASEAN talks over Kampuchea. In ideological terms Burma is far from easy to categorise. Since 1962 the 'Burmese road to socialism' has rested on a rather eclectic selection of Marxist and Buddist philosophy (Steinberg, 1981).

These groupings are, however, far from fixed. Changes in the world-economy are giving rise to new structures that may radically alter or supersede them. Is Singapore's status as a newly industrialising country more significant than location in South East Asia or membership of ASEAN? Would the city state be better examined with Hong Kong, South Korea and Taiwan as one of the 'four little tigers'? Should South East Asia be considered part of a wider Asian economy, part of the increasingly advocated Pacific Rim or as part of the globe passing increasingly under Japanese hegemony?[3]

The present book does not set out to develop specifically regionally orientated theory; rather the intention is to provide a broad study of South East Asia from a world-economy perspective. Thus the present study begins with a review of South East Asia's role in the world-economy, outlines the developments that give rise to the present pattern and isolates the key elements of contemporary change. Given this focus the account of South East Asia since the 1940s is concerned almost entirely with the ASEAN group. An analysis of the complex and distinctive experiences of Burma, Laos, Kampuchea and Vietnam are beyond the scope of the present volume. Similarly for the ASEAN group no attempt is made at a comprehensive review of their post-war development. Given that the main concern is the interaction with the world-economy, emphasis is placed on those states, sectors and strategies which illustrated the process most clearly.

Since the early 1970s South East Asia has entered a new and perhaps critical stage of development. This has been characterised by rapid development of manufacturing sectors, the emergence of authoritative, corporatist regimes and complex bureaucratic structures. Essentially this is a result of the acceleration of the process of internationalisation, the key element in the restructuring of capital since the late 1960s, and the region's significant role in the New International Division of Labour.

The ASEAN states have become a major focus for international capitalism. Devan (1987: 159) has described South East Asia as 'one of the most vibrant regions in the world'. Attention is drawn to the region's resource endowment; market potential, with a population of over 300 million and some of the higher per capita incomes in the Third World (table 1.1); possibilities for investment given low labour costs (table 1.2) and comparative political stability; strategic position between the Indian and Pacific oceans; and the 'openness' of the economies to trade, foreign investment and the activities of multinational corporations. This interest in the region by the agents of international capital has been reinforced by the maintenance of comparatively high rates of economic growth during the uncertain international conditions that have prevailed since the early 1970s (table 1.3). To a considerable degree, the high rates of growth have been a reflection of the favourable appraisal of the region by investors.

There has been considerable debate over South East Asia's resource endowment and consequent importance to the world economy (Fisher 1964; Fryer 1979). However, the development of primary production and specialisation during the colonial period has dictated that South East Asia remains a significant producer of the limited range of strategic commodities (table 1.4).[4] For others, notably rice and cassava, the

Table 1.1 *South East Asian economies in perspective, 1986*

	GNP per capita (US $)	World Bank income group	Rank[1] income group	Life expectancy	Infant mortality (1000 live births)	Crude birth rate (1000 population)	Crude death rate (1000 population)	Daily calorie intake
Burma	200	Low	9	59	70	33	10	2508
Indonesia	490	Lower-middle	36	57	84	28	11	2476
Kampuchea	n/a	n/a	n/a	n/a	160	n/a	n/a	2171
Laos	159[2]	n/a	n/a	50	122	39	15	2317
Malaysia	1850	Upper-middle	68	69	30	29	6	2601
Philippines	560	Lower-middle	38	63	51	35	7	2260
Singapore	7410	Upper-middle	87	73	9	16	5	2696
Thailand	810	Lower-middle	49	64	48	25	7	2399
Vietnam	n/a	n/a	n/a	65	59	34	7	2281
Low Income	270			61	99	30	10	2329
Lower Middle Income	750			59	95	35	9	2719
Upper Middle Income	1890			67	65	27	8	2967

[1] For those countries where data is available.
[2] Far Eastern Economic Review *Yearbook*, 1988.
Source: World Bank, *World development report* 1988: United Nations, 1987

Table 1.2 *The attraction of South East Asia for labour–intensive manufacturing investment*

The Coats Patons table of comparative labour costs at 21 April 1981

	Single shift		Double shift		Treble shift	
	Total cost per hour (£)	Index	Total cost per hour (£)	Index	Total cost per hour (£)	Index
UK (base)	2.678	100	3.186	100	3.481	100
Italy	3.259	122	3.499	110	4.943	121
West Germany	3.561	133	3.696	116	3.913	115
Canada	3.596	134	3.564	112	3.613	109
United States	3.134	117	3.134	98	3.157	90
Portugal	1.076	40	1.177	37	1.799	42
Columbia	0.950	36	1.121	35	1.304	26
Brazil	0.840	31	1.009	32	1.065	31
Peru	0.611	23	0.620	19	0.637	19
India	0.342	13	0.345	11	0.416	11
Philippines	0.276	10	0.276	9	0.282	8
Indonesia	0.166	6	0.169	5	0.168	5

Source: *The Financial Times*, 29 June 1981, 11

Table 1.3 *Growth of GDP, average annual percentage rate*

	1963–73	1974–80	1980–86
Burma	2.9	6.8	4.9
Indonesia	8.2	7.1	3.4
Malaysia	9.4	7.7	4.8
Philippines	6.0	6.3	−1.0
Singapore	13.0	8.2	5.3
Thailand	6.7	7.6	4.8
Hongkong	10.2	8.6	6.0
South Korea	10.8	9.9	8.2
Taiwan	11.0	9.8	6.4
Low income	5.6	4.6	7.5
Lower middle income }	7.4	4.4	1.8
Upper middle income }			2.5
Third World	6.5	5.4	3.8
Industrial market economies	4.5	2.3	2.5

Source: World Bank, *World development report* 1983 and 1988

region is a major source of exports. Similarly, for some developed countries, South East Asia is a major source of key raw materials. Japan, for example, is dependent on Indonesian oil and liquid natural gas.[5] The growth of the service sector, to underpin industrial development, supply the needs of the local population and earn foreign exchange, has increasingly attracted foreign investment. In addition the tourist industry has attracted increasing amounts of foreign capital as well as becoming a major earner of foreign exchange (Richter, 1986). Further, since the early 1970s the region has become an important source of labour exports, particularly from Thailand and the Philippines to the Middle East (ESCAP, 1987).

Table 1.4(a) *South East Asian primary production, 1986*

	million metric tons	% of world production	% of world trade
Cassava	38.00	30.4	76.3
Coconuts	20.01	64.7	47.9
Copra	3.01	74.7	43.6
Palm kernels	1.26	52.2	75.4
Palm oil	4.91	70.9	81.8
Rubber	3.39	77.8	88.3
Tin concentrates	69094.00	58.4	68.0
Rice	100.64	21.2	24.3

[1] Metric tons of concentrate

Table 1.4(b) *National percentage contributions to South East Asia primary production and export, 1986*

	Oil		Tin		Rubber		Oil palm		Cassava	
	Production	Export[1]	Production	Export	Production	Export	Production	Export	Production	Export
Brunei	8.2	7.9	2.5	—	—	—	—	—	—	—
Burma	0.8	—	—	—	—	—	—	—	—	—
Indonesia	59.4	57.7	28.8	—	32.5	28.9	21.7	13.5	35.1	—
Malaysia	29.2	32.6	47.3	75.3	42.3	45.7	75.9	85.2	1.0	—
Philippines	0.3	0.5	—	—	2.4	0.4	0.4	0.2	4.5	—
Thailand	2.2	1.3	21.3	24.7	21.1	24.7	2.0	1.0	51.5	100.0
Vietnam	[2]	—	—	—	1.6	—	—	—	7.9	—

	Coconuts		Copra		Palm kernels		Rice	
	Production	Export	Production	Export	Production	Export	Production	Export
Brunei	—	—	—	—	—	—	—	—
Burma	0.6	—	—	—	—	—	13.8	—
Indonesia	44.7	0.3	—	—	2.0	—	40.5	—
Malaysia	6.7	39.3	34.4	23.5	95.5	86.4	—	—
Philippines	40.1	59.2	61.7	71.4	0.7	11.1	8.6	—
Thailand	4.9	1.2	0.9	—	1.8	2.5	20.0	100.0
Vietnam	2.5	—	3.0	5.1	—	—	15.1	—

[1] Share of the value of petroleum and petroleum products exported by producers of crude oil. Excludes re-exports from Singapore.
[2] Offshore oil production started late in 1986.
Source: *Agricultural Commodities*, FAO 1986; United Nations, *International trade yearbook*, Geneva, 1988.

Access to these resources, markets and cheap labour remains of major importance to the industrial nations, most significantly the USA and Japan. The need to safeguard these interests is inextricably bound up with the strategic position South East Asia has come to occupy in the global and Western Pacific balance of power.

1.2 South East Asia and the balance of power

The importance of South East Asia to the world-economy has resulted in the region becoming a focus of conflict, initially of colonial rivalries, later of nationalist struggles and most significantly of the conflict between imperialism and the forces of social revolution. American defeat and withdrawal from Vietnam in 1975 has not diminished South East Asia's focal position in the struggle between imperialism and the forces of social revolution at the global and, more specifically, Western Pacific scales. The region can only be understood against the background of its key role in the military and economic strategy of the USSR, China, Japan and the USA. Crucial to our understanding are the policies and attitudes that the occupation of this strategic location has engendered in the regions' states.

The formation in 1967 of ASEAN (Association of South East Asian Nations), comprising Indonesia, Malaysia, Philippines, Singapore and Thailand (Brunei became a full member in 1984), marked an important step in the establishment of an organised pro-American bloc in South East Asia. While ostensibly ASEAN originated within the region – Adam Malik, the Indonesian foreign minister, is credited with originating it – and was an economic and cultural organisation, there is little doubt the driving force was American and the aims political and military (Caldwell, 1978: 14). Limited progress has been made in establishing economic cooperation and integrating the markets of member states. However, in military and political affairs developments have been considerable, particularly since the reunification of Vietnam in 1976. While there are considerable similarities between the political structures and complexions of the ASEAN states, and for some purposes they have come to act as a unit (for example in some trade negotiations), the relationship between the member nations remains uneasy. A number of long-standing international disputes (for example, the ownership of East Timor), mostly legacies of the colonial period, remain unsettled but largely dormant. However, Indonesia's closure in 1988 of the Sunda Straits to international shipping illustrates the potential for intra-regional conflicts (Leifer, 1989). In recent years the exploitation of off-shore oil and gas reserves has produced dissension amongst member states (Lee, 1979; Siddayao, 1978; Valencia, 1986). In addition the economies are in increasing competition to obtain markets and attract foreign investment and multinational corporations. The joint approaches, for example to the European Community over textile quotas, have on occasions been undermined by members 'breaking ranks' and establishing a short-term gain through bilateral agreement.

The ASEAN states are superficially unified in their anti-communist and generally anti-Vietnam stance. In this the central issue has become the resolution of the Kampuchean 'problem'. However, they have shown themselves far from agreed on the nature of the 'solution', how best to achieve it or, indeed, their relations with the 'socialist' countries as a whole (Buss, 1985; 176: 8). There are, for example, considerable differences in the attitudes towards China. While, to a degree, China is perceived as a threat, its anti-USSR and anti-Vietnamese position fits into the policies of the ASEAN states. Thailand has in recent years developed much closer trading links with China, and this has been met with disapproval from Indonesia and Malaysia.[6] To a degree, these states still regard China with suspicion because of past links with insurgency movements. In addition, Thailand has been prepared to negotiate trade agreements with USSR and the Eastern European states.

Similarly Malaysia, while subscribing to the general anti-Vietnamese position, perceived that the end of the war would give rise to a wide variety of economic opportunities. Between 1978 and 1982 Malaysia and Vietnam agreed a series of joint projects; this created considerable friction with the other ASEAN members, particularly Thailand.[7]

Primarily, South East Asia has to be seen in the context of American strategy in the Western Pacific. While Japan remains the lynch-pin of American defence policy in the region major secondary roles are played by South Korea, Taiwan and the Philippines. From the American viewpoint:

> The real gap on the West Pacific rim is Vietnam with its most sensitive and emotional experience and which remains of great strategic importance ... the question over Vietnam's position is serious enough to make it a prime concern in American policymaking. Vietnam appears to the true unfinished business in the sino-American rapprochement. On the other hand, the problem of its relations either with the Soviet Union or with China is of decisive significance for the entire strategic set-up in East Asia and, on the other hand, the extent of its Indochinese hegemony determines the policy of the ASEAN states. (Bartalits and Schneider, 1987: 44)

This preoccupation with Vietnam has been fueled by Vietnam's membership of CMEA (Council for Mutual Economic Assistance), the 1978 treaty with the USSR, the establishment of 'sovereignty' over Laos, the

Table 1.5 *Defence expenditure, 1986*

	As a % of GNP	As a % of Government expenditure
Brunei	10.0	32.4
Burma	3.0	21.4
Indonesia	2.5	9.3
Malaysia	4.6	16.6
Philippines	1.3	11.9
Singapore	6.3	22.5
Thailand	4.4	30.8
South Korea	5.3	30.6
Taiwan	8.2	35.0
United Kingdom	2.0	16.7
USA	7.9	32.2
Low income group	3.7	17.7
Lower middle income	3.9	15.8
Upper middle income	2.9	10.3
Third World	3.3	14.3
Industrial market economics	4.7	28.6

Source: *Far Eastern Economic Review Year Book*, 1988 and World Bank, *World development report*, 1988.

invasion of Kampuchea and, above all, by the leasing of naval facilities at the ports of Danang and Cam Ranh to the USSR. The American view that Vietnam and the USSR are 'threats' to regional and global peace permeates ASEAN. Member states express fears of support for insurgency movements, border clashes and invasion in the case of Thailand,[8] and of Soviet sea power based in Vietnam dominating the sea lanes.[9] Whatever the veracity of these views there is little doubt that they dominated strategic thinking in ASEAN. In addition, for the ruling classes of these states, Vietnam, whatever its problems, remains a powerful reminder that there is an alternative to accommodation to international capital.

The South East Asian states have remarkably high levels of defence expenditure. The figures contained in table 1.5 seriously under represent the level of expenditure, particularly in the Philippines and Thailand, because of the exclusion of American military 'aid'. Mirza (1986: 47–9) has described Singapore as a 'garrison state', and this term could equally well be applied to all the ASEAN countries.

While this high level of defence expenditure is closely related to the strategic position of the region, internal security remains a major consideration for all the states. This is still frequently presented in terms of the continued threat of 'communist subversion', often merely a euphemism for any opposition to the government. Indeed since independence the region has been characterised by repressive, often military or military-backed regimes. Only in Malaysia has there been no direct military intervention in the government.[10] Caldwell (1978: 14), with some justification, has described the members of ASEAN as 'possessing impeccable reactionary credentials'. While the repressive state apparatus may well ensure a stable environment for the international capital it is extremely costly. The World Bank (*World Development Report*, 1988: 106–7), has concluded that:

> evidence increasingly points to high military spending as contributing to fiscal and debt crisis, complicating stabilization and adjustment and negatively affecting economic growth and development.

The Soviet presence at Cam Ranh and the entrenched American base structure in the Philippines (Clark Air Base and Subic Bay Naval Base) are the clearest indicators of the vital position that South East Asia has come to occupy in global affairs.

Japanese influence in South East Asia is steadily increasing and it can be argued that the ASEAN nations are rapidly becoming part of a Japanese-dominated economic sphere. The increasing importance of Japanese trade and investment is discussed in sections 1.3 and 1.4. This development is likely to have serious implications for Japanese–American relations.

Despite the strategic and economic importance of the ASEAN group the overall balance of power in the Western Pacific is a result of the interplay between the USSR, China, Japan and the USA; the states of South East Asia have little say in this (Tjeng, 1985: 190).

1.3 Foreign direct investment

Since the mid-1960s Third World countries have become increasingly important destinations for f.d.i. (foreign direct investment), their global share rising from 18 per cent in 1965 to 30 per cent in 1986. This investment is, however, heavily concentrated in a small number of the higher-income countries. In 1983, Brazil (US$ 24.6 billion), South Africa (US$ 17.1 billion), Mexico (US$ 25.0 billion), Argentina (US$ 5.8 billion) and ASEAN (US$ 25.0 billion) accounted for almost 60 per cent of f.d.i. flowing to the Third World.

The ASEAN states have increased their share of Third World f.d.i. from 10.2 per cent in 1973 to 16.9 in 1986. However, the region's stock of f.d.i. is heavily concentrated in Singapore (45.9 per cent) and Malaysia (32.7 per cent). As may be seen from table 1.6 this concentra-

Table 1.6 *Percentage share of net ASEAN foreign investment*

	1965–71	1972–76	1977–80	1981–86
Indonesia	21.7	17.3	12.1	11.6
Malaysia	31.2	27.5	31.6	35.7
Philippines	–[1]	4.8	2.8	2.7
Singapore	27.6	42.5	48.2	50.0
Thailand	19.5	7.6	5.3	8.9

[1] Between 1965 and 1971 the stock of foreign investment in the Philippines fell by US$ 58.1m.
Source: IMF Balance of Payments Yearbook, various issues

tion has increased markedly since the mid-1960s, with these countries' f.d.i. showing exceptionally high rates of growth. In contrast the Philippines and, until recently, Thailand have been far less attractive to foreign investors.

Singapore's dependence on foreign investment is without parallel. Mirza (1986: 2–3), with much justification, describes Singapore as the most internationalised economy in the world. In 1983 foreign corporations and residents contributed over 25 per cent of the country's GDP, and 31 per cent of gross domestic capital formation. This dependence, while integral to the high rates of growth that Singapore has experienced (table 1.3) has also made the city state extremely vulnerable to changes in the world-economy (see chapter 5).

While the level and global share of the ASEAN states' f.d.i. increased rapidly until the early 1980s the growth has been far from even (figure 1.1). The variation in flows naturally reflect the general climate for the world-economy and the perception by investors of the region's and individual states' long-term stability and profitability. All the ASEAN countries have taken broadly similar steps to encourage f.d.i. (see chapter 5). However, under prevailing economic conditions the incentive schemes of the individual states can have only a marginal effect on foreign investment decisions.

Between 1981 and 1986 there was major disinvestment throughout the region, most significantly in Thailand, Malaysia and, above all, the Philippines. Foreign interests have been sold to local concerns, for example the sale of Dunlop's Malaysian rubber plantations and facilities in 1986.

For investment, the question of political stability is paramount. Despite insurgency and occasional violent protests, some of the ASEAN nations have in recent years been characterised by long-lasting regimes – Lee in Singapore, Suharto in Indonesia and Marcos in the Philippines.

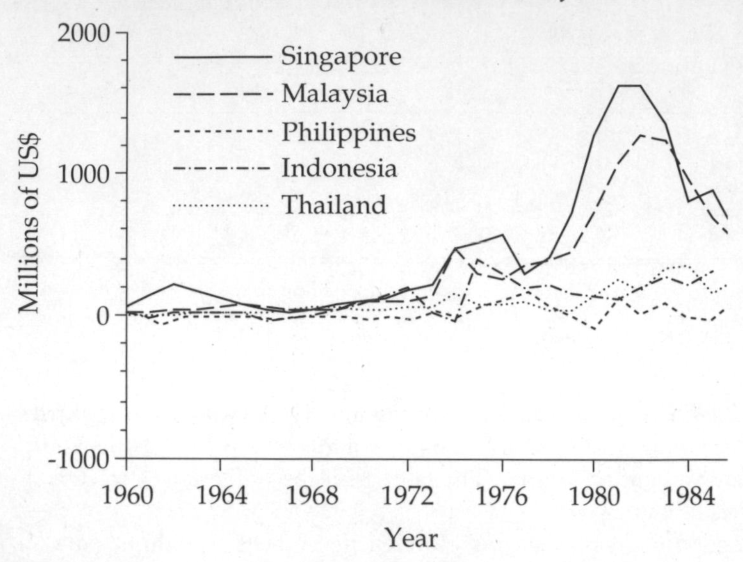

Figure 1.1 Annual net inflows of foreign direct investment, 1960–86
Notes:
Malaysia, 1960–2 refers to the Federation of Malaysia;
1965–6 include IBRD loans to the National Electricity Board.
Singapore, 1960–8 include some IBRD loans to the Public Utilities Board and the Port of Singapore.
Source: IMF *Balance of Payments Yearbook*, Washington, various issues

Only in Thailand have governments been short lived, and investors seem to be accepting that even when the change of government involved a *coup d'état* there is little threat to investment.

It is important to stress that investment risk assessors' perception of South East Asia as politically stable is coloured by the apparent instability of much of Africa and Latin America. In the search for a cheap, subservient labour force, accommodation to the interests of international capital and security of investment the ASEAN countries have much to recommend them. It would be naive to assume that foreign investors are unaware of the possibilities of instability, if not violent political change in South East Asia. However, investment is increasingly short-term and mobile, the risks are perceived as greater elsewhere, and it has become necessary to accept high levels of risk. The ephemeral nature of much Third World f.d.i. is illustrated by the rapid outflow of funds from the Philippines following the murder of Aquino in 1983, and the sharp influx after 1986.

Shorter term and far less spectacular ebbs and flows of f.d.i. have occurred elsewhere in the region since 1981. Much capital, particularly

Asian, has found outlets in the region's burgeoning stock and commodity exchanges, property speculation and the development of the service sector. Additionally short-term financial flows have moved closely with differential interest rates. In all of this South East Asia is both a reflection and part of global movements in a period of continued economic and political uncertainty.

The net inflow of f.d.i. does not give the whole picture. Many foreign investors choose to raise money on the domestic money markets. This is particularly the case in the Philippines and Thailand (see chapter 5). International funds flowing into these markets are channelled into a variety of foreign-owned undertakings, thus substantially reducing their risks. On the other side of the equation substantial funds are moved out. The outflow of funds from Singapore has, for example, resulted in the city state's investors becoming major holders of American real estate. In addition a considerable volume of funds is believed to move out of the region illegally. This whole area is, for obvious reasons, poorly documented. In the case of the Philippines, it is variously estimated that between 1974 and 1983 US$ 4.2 to 5.9 million left the country illegally.[11] Indeed from the early 1970s until the murder of Aquino in 1982 the illegal outflow greatly exceeded the inflow of investment (Jayasuriya, 1987: 96).

Considerable confusion surrounds both the origins of South East Asian f.d.i. and its sectoral distribution (Ariff and Hill, 1985: 30–34; Chia Siow Yue, 1986; Hill, 1988; Hill and Johns, 1985; Naya, 1987: 69–79, 69;[12] Thus the data contained in tables 1.7 and 1.8 should be regarded as indicative only. However, since 1970 the pattern of South East Asian f.d.i. has changed markedly with, in particular, Japan displaying European and, to a lesser extent, American sources. Additionally there have been increased inflows from other Third World countries, particularly Hongkong, South Korea and Taiwan, as well as significant transfers of funds within ASEAN. The USA however, remains the largest single source of f.d.i. for Indonesia, the Philippines, and, probably, Thailand (table 1.7).

For ASEAN as a whole the bulk of f.d.i. has in recent years flowed into manufacturing and the service sector (table 1.8). The service sector is emerging as the region's fastest growing sector with, in particular, Japanese capital moving into transport, trade, restaurants, stores and hotels. The growth of f.d.i. has had substantial implications not only for the sectorial pattern of development of the ASEAN countries but also for the spatial. This is well illustrated in the case of Indonesia where the concentration of manufacturing f.d.i. in Jakarta, West Java and, to a lesser extent, Surabaya, reinforces the established Indonesian 'core-periphery' structure (see chapter 4 and 5).

Table 1.7 *Source of South East Asian foreign direct investment, percentage share*

	Indonesia	Malaysia	Philippines	Singapore	Thailand
	Realised foreign investment, 1967–85 %	Fixed assets as at end 1980 %	Approved investment 1968–80 %	Fixed assets as at end 1979 %	Registered capital 1960–81 %
Japan	21.0	19.6	22.5	16.5	31.3
United States	58.0	10.7	29.8	28.6	11.1
Europe	13.0	28.1	27.8	36.1	14.0
Other ASEAN }		23.9	0.3	n/a	5.3
North East Asian[1] }	5.0	9.3	2.9	9.5	19.2
Other	3.0	8.6	16.7	9.5	19.2
Total	100.0	100.0	100.0	100.0	100.0

Notes: Country totals not directly comparable because they refer to different time periods, measures and coverage. The data for Singapore and Malaysia refer only to manufacturing, services and excluded from the Philippines.

[1] Refers to Hong Kong, Korea and Taiwan. Taiwan not included in the case of the Philippines. S. Korea not included in the case of Singapore. See also Naya (1987: 73) who asserts that the USA remains the largest foreign investor in Thailand.

Source: Indonesia, Hill, 1987: 55; Malaysia, Malaysian Development Authority, cited Ariff and Hill, 1985: 31; Philippines, von Kirchbach, 1982: 14, cited Ariff and Hill, 1985: 31; Singapore, Chia, 1982: 286, cited Ariff and Hill, 1985: 31; Thailand, Board of Investment.

Table 1.8 *Estimated sectorial distribution of foreign direct investment, percentage share*

	Indonesia			Malaysia			Philippines			Singapore			Thailand		
Sector	1967–70	1971–76	1977–83	1966–70	1971–76	1977–80	1970	1970–76	1977–83	1970	1971–76	1977–82	1972	1973–76	1977–83
Agriculture, forestry, and fishery	14.1	4.7	1.3	34.6	27.3	18.6	0.0	1.3	0.9	0.0	0.0	0.0	0.3	0.3	0.3
Mining and petroleum	64.8	64.8	88.3	12.4	12.3	11.9	14.5	11.2	22.3	0.0	0.0	0.0	12.0	15.6	17.3
Manufacturing	17.4	25.8	9.5	21.3	25.1	29.6	57.7	48.7	35.7	75.6	68.4	64.4	21.6	30.5	34.2
Food, beverages, and tobacco	4.8	1.8	0.4	4.8	2.1	3.1	14.8	4.5	8.2	2.3	2.5	2.1	0.4	3.3	1.2
Chemical products	4.1	3.4	2.8	2.8	2.0	1.2	7.2	7.4	11.6	4.6	3.5	2.5	4.8	4.6	3.7
Metal products	4.0	5.3	2.9	2.7	0.8	2.0	4.0	14.8	3.9	4.0	2.2	2.8	0.2	4.1	5.3
Machinery and transport equip.	n/a	n/a	n/a	0.4	0.3	0.3	2.3	1.6	2.9	3.9	13.3	7.7	1.9	0.8	4.2
Electrical products	n/a	n/a	n/a	1.0	5.6	4.9	2.5	n/a	n/a	6.2	8.2	10.4	3.2	4.4	10.5
Textiles	2.9	11.0	2.0	1.3	5.4	2.2	1.9	3.3	1.6	3.4	4.6	3.7	12.5	13.2	3.4
Wood and pulp	0.1	0.6	0.4	0.4	0.8	0.1	5.9	n/a	n/a	2.6	4.0	3.0	n/a	n/a	n/a
Petroleum and coal	n/a	n/a	n/a	5.7	2.7	6.9	n/a	5.5	0.4	42.2	24.1	27.0	n/a	n/a	n/a
Others	1.6	3.4	1.0	2.1	5.4	8.9	19.0	11.6	7.1	4.9	6.5	5.2	3.4	0.1	1.6
Services	3.4	4.2	0.7	17.0	26.8	34.3	27.4	38.3	15.0	24.4	31.5	36.5	44.0	47.4	31.5
Construction	0.3	0.5	0.1	0.8	1.0	1.3	0.3	0.1	1.0	n/a	n/a	n/a	22.1	6.3	15.9
Others	0.0	0.0	0.1	13.9	7.5	4.5	0.1	0.1	0.0	n/a	n/a	n/a	0.0	0.0	0.0
Total	100.0	100.0	100.0	100.0	100.0	100.0	100.0	100.0	100.0	100.0	100.0	100.0	100.0	100.0	100.0

Note: These figures are based on a wide variety of national sources and should be treated with caution.
Source: Pangestu, 1985 cited Naya, 1987: 74

1.4 Trade

South East Asian countries maintained comparatively high rates of trade growth during the 1970s and 1980s (table 1.9). The region increased its share of Third World foreign trade from 4.6 per cent in 1978 to 6.4 in 1986 and its share of less-developed Asian trade (including China) from 32 to 36 per cent. This comparatively high rate of growth is a reflection of the region's increased integration into the world-economy.

As may be seen from table 1.10 the South East Asian economies are by no means uniformly integrated into the international economy. Burma, despite limited moves since the early 1970s to 're-engage', shows little sign of increased integration. In contrast Singapore stands out as by far the most integrated. Between these extremes Malaysia is on a par with the East Asian Newly Industrialising Countries.

As with f.d.i. the increased importance of South East Asia in the world-economy has been accompanied by shifts in the direction of trade. The position of the EC countries has been eroded by the rapid growth of Japanese influence.[13] Japan is now the dominant trading partner with the USA playing a still strong but, very clearly, secondary role (table 1.11). Again, as with f.d.i., there are marked differences between countries with, for example, the USA remaining dominant in the Philippines.

The composition of South East Asian trade has changed dramatically over the last twenty years (table 1.12). While exports are still dominated by primary production, outside of the socialist bloc the export of manufactured goods has become increasingly important.[14] The decline in the relative importance of primary exports has been exaggerated by steep falls in commodity prices since 1979 (figure 1.2). To a degree the expansion of the manufacturing sectors, whether for the domestic or export market, has been accompanied by increased imports of components, raw materials and capital goods. Indeed, by 1986, 60 per cent of the value of ASEAN imports fell into these categories. The change in the composition of trade has not taken place uniformly between the region's major trading partners. In 1986 around 50 per cent of South East Asian exports to the EEC and USA were manufactured goods, compared with around 25 per cent in 1973. This contrasts with exports to Japan, 95 per cent of which were raw materials. It can be argued that the Japanese role in South East Asian trade and f.d.i., has a strong 'neo-colonial' flavour.

Intra-regional trade remains limited despite the efforts directed towards this since ASEAN was established in 1967.[15] The 1977 ASEAN

Table 1.9 *Growth of merchandise trade, 1960–86 (percentage annual average rate of growth)*

	Exports			Imports		
	1960–70	1970–80	1980–86	1960–70	1970–80	1980–86
Burma	−11.6	1.3	−8.8	−5.7	−2.8	−8.8
Indonesia	3.4	6.5	2.0	2.0	11.9	−1.0
Malaysia	5.8	6.8	10.2	2.3	7.1	5.2
Philippines	2.2	7.7	−1.7	7.2	2.6	−6.0
Singapore	4.2	12.0	6.1	5.9	9.9	3.6
Thailand	5.2	11.8	9.2	11.4	4.9	2.0
Hongkong	12.7	9.7	10.7	9.2	12.1	7.9
South Korea	33.4	22.0	13.1	20.6	10.9	9.3
Low income	4.9	−0.7	6.5	5.3	2.4	7.2
Lower middle income	5.2	3.0	2.4	6.5	4.1	−2.4
Upper middle income	5.4	7.0	5.6	5.9	4.7	−0.1
Industrial market economies	8.5	5.4	3.3	9.5	4.4	4.3

Source: World Bank, *World development report* 1983; 1986; 1988

Table 1.10 *Integration into the world economy: the value of foreign trade as a percentage of GNP*

	1972	1985
Burma	14.9	15.5
Indonesia	29.8	45.6
Malaysia	70.1	90.6
Philippines	45.0	39.5
Thailand	27.6	44.1
Singapore	198.1	277.0
Hongkong	156.4	175.0
South Korea	40.9	73.9
Taiwan	71.5	80.5

Source: World Bank, *World development report* 1983 and 1987

Preferential Trading Arrangements introduced tariff preferences and paved the way for a gradual reduction of trade barriers between member states. In addition it was envisaged that increased specialisation and a clear regional division of labour would emerge (Kraus and Lütkenhorst, 1986: 36). While progress in the number of preferred items has been impressive – perhaps 20,000 by 1986,[16] few of these have been significant trade items and effective trade liberalisation has been extremely limited (Tan, 1982). While there has been a general increase in the volume of trade between ASEAN members there has been little change in its relative importance (table 1.13).

Table 1.11(a) *Export intensities of South East Asian countries with Japan, U.S.A., European Community and ASEAN, 1984*

Exporting country i	Importing country j	Japan	U.S.A.	European Community	ASEAN
Indonesia		6.4	1.1	0.1	2.8
Malaysia		3.0	0.7	0.4	6.6
Philippines		2.7	2.1	0.4	2.5
Singapore		1.3	1.1	0.3	7.3
Thailand		1.8	0.9	0.7	3.5

Calculated as: $Ix = \frac{Xij}{Xi} : \frac{Mj}{Mw - Mi}$

Ix = Intensity of exports of i to j
Xij = Exports of i to j
Xi = Total exports of i
Mj = Total imports of j
Mw = Total world imports
Mi = Total imports of i

Table 1.11(b) *Import intensities of South East Asian countries with Japan, U.S.A., European Community and ASEAN, 1984*

Importing country i	Exporting country j	Japan	U.S.A.	European Community	ASEAN
Indonesia		2.5	1.5	0.4	3.3
Malaysia		2.7	1.3	0.4	4.9
Philippines		1.4	2.2	0.3	2.7
Singapore		1.4	1.2	0.2	6.0
Thailand		2.8	1.1	0.4	3.3

Calculated as: $Im = \frac{Mij}{Mi} : \frac{Xj}{Xw - Xi}$

Im = Intensity of imports of i to j
Mij = Imports of i to j
Mi = Total imports of i
Xj = Total exports of j
Xw = Total world exports
Xi = Total exports of i

Source: Kraus and Lütkenhorst, 1986: 22, calculated from data of IMF, *Direction of Trade Statistics Yearbook*, 1985, Council for Economic Planning and Development, Republic of China, *Taiwan Statistical Data Book*, 1985

1.5 South East Asian economies and societies

In terms of the measures of development generally used, the capitalist states of South East Asia are regarded as amongst the least impoverished in the Third World (table 1.1). Comparable data on the socialist bloc is scant, often dubious and not necessarily meaningful. It would be surprising, given their recent history, if these states were not experiencing serious economic problems. According to World Bank statistics, however, on a number of measures – infant mortality rate, daily calorie intake, crude death rate and life expectancy – lower figures are found in some of the pro-capitalist economies.

Within the capitalist group there is remarkable variation. At one extreme is Singapore, which, on many indicators, ranks with the developed countries (table 1.1). Indeed, there is mounting pressure by OECD members to remove the city-state's 'less-developed' status. The contrast between Singapore and Malaysia – again on these measures one of the most developed countries in the Third World – is striking. So too is the difference between Malaysia and the other states. At the level of these crude indicators, South East Asia is a region of starkly uneven development. Over the last twenty years the disparities between the region's nations has tended to increase.[17]

Table 1.12 *Changing composition of the value of international trade, 1965–1986*

Imports – percentage by value

	Food			Fuels			Other primary products			Machinery and transport equipment			Other manufactured goods		
	1965	1986	Change	1965	1986	Change	1965	1986	Change	1965	1986	Change	1965	1986	Change
Burma	15	6	−9	4	1	−3	5	2	−3	18	43	+25	58	48	−10
Indonesia	6	4	−2	3	14	+11	2	4	+2	39	39	0	50	38	−12
Malaysia	25	10	−15	12	5	−7	10	4	−6	22	51	+29	32	30	−2
Philippines	20	8	−12	10	15	+5	7	5	−2	33	22	−11	30	51	+21
Singapore	23	9	−14	13	20	+7	19	5	−14	14	37	+23	30	30	0
Thailand	6	5	−1	9	12	+3	6	8	+2	31	34	+3	49	40	−9

Exports – percentage by value

	Fuels, minerals and metals			Other primary products			Machinery and transport equipment			Other manufactured goods		
	1965	1986	Change	1965	1986	Change	1965	1986	Change	1965	1986	Change
Burma	5	3	−2	94	84	−10	0	9	+9	0	4	+4
Indonesia	43	58	+15	53	21	−32	3	3	0	1	19	+18
Malaysia	35	26	−9	59	38	−21	2	26	+24	4	10	+6
Philippines	11	14	+3	84	26	−58	0	6	+6	6	55	+49
Singapore	21	21	0	44	12	−32	11	38	+27	24	30	+6
Thailand	11	4	−8	84	56	−28	0	9	+9	4	33	+29

Source: World Bank, 1988

Table 1.13 *Intra-ASEAN exports as a percentage of the total value of exports, 1970–87*

Exporting countries		Brunei	Indonesia	Malaysia	Philippines[1]	Singapore	Thailand	Total
Brunei	1970	-	0	82.2	0	1.0	0	83.2
	1976	-	0	6.3	1.7	1.7	0	9.7
	1982	-	0	0.03	2.2	0.4	0.5	3.1
	1987	-	0.03	0.6	2.6	6.6	12.3	21.6
Indonesia	1970	0	-	3.2	2.3	15.5	0	21.0
	1976	0	-	0.3	1.1	7.5	0.02	8.9
	1982	0	-	0.3	1.0	5.2	0.02	6.52
	1987	0.05	-	0.6	0.4	6.1	0.6	7.8
Malaysia	1970	0.6	0.6	-	1.7	21.6	0.9	25.4
	1976	0.7	0.4	-	1.5	18.3	1.3	22.2
	1982	0.3	0.3	-	1.0	25.7	3.0	30.3
	1987	0.5	0.8	-	1.8	18.0	2.8	23.9
Philippines	1970	0	0.2	0	-	0.7	0.3	1.2
	1976	0	0.5	0.2	-	2.2	0.3	3.2
	1982	0.1	1.1	3.6	-	2.2	0.3	7.3
	1987	0.8	1.1	2.0	-	3.4	21.7	29.2
Singapore	1970	1.6	n.a.	21.9	0.3	-	3.3	27.1
	1976	1.5	n.a.	15.2	0.8	-	3.0	20.8
	1982	1.7	n.a.	17.7	1.6	-	3.8	24.8
	1987	1.2	n.a.	14.2	1.5	-	4.2	21.1
Thailand	1970	0	2.3	5.6	0.1	7.0	-	15.0
	1976	0.1	5.2	4.2	1.0	6.8	-	17.2
	1982	0.1	2.0	5.4	0.5	6.8	-	14.7
	1987	0.1	3.3	3.1	0.6	8.7	-	15.8

Total value of intra-ASEAN exports as a percentage of the total value of members exports		
	1970	21.5
	1976	14.8
	1982	22.7
	1987	17.1

[1] Philippine exports to Thailand have grown sharply since 1984

Source: *IMF Directory of Trade Statistics Year Book*, Washington, various issues

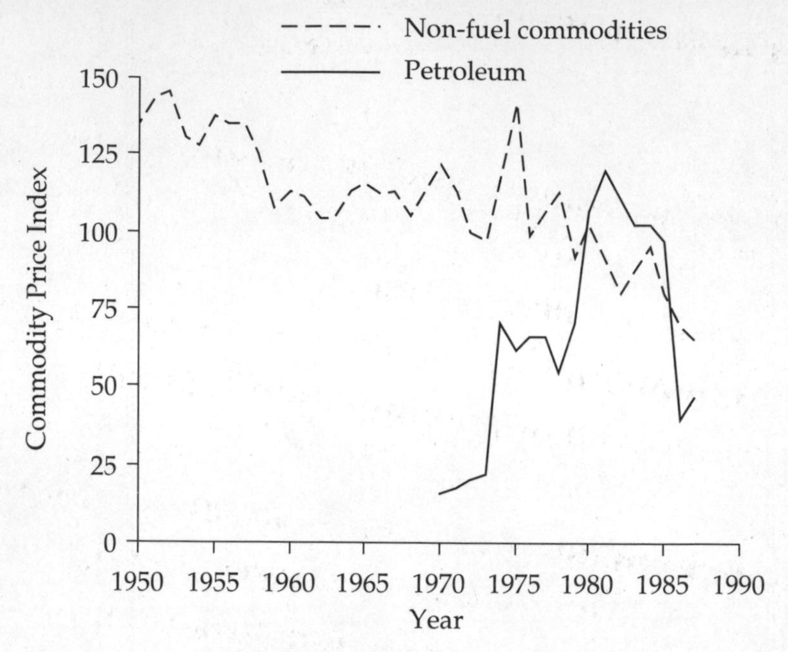

Figure 1.2 Commodity prices 1950–87
Notes: The index measures commodity prices in terms of the price of manufactures imported by developing countries. The non-fuel commodities are coffee, cocoa, tea, maize, rice, wheat, sorghum, soybeans, groundnuts, palm oil, coconut oil, copra, groundnut oil, soybean meal, sugar, beef, bananas, oranges, cotton, jute, rubber, tobacco, logs, copper, tin, nickel, bauxite, aluminium, iron ore, manganese ore, lead, zinc and phosphate rock.
Source: World Bank, *World Development Report*, Washington, 1983 and 1988

While, apart from Singapore, the majority of the South East Asian population lives in rural areas and is dependent on agriculture, the internal structures of the economies are extremely varied (table 1.14). Generally, since the mid-1960s, there has been some convergence in the contributions of the industrial, agricultural and service sectors to GDP. However, these broad categories conceal considerable variation. In Indonesia, for example the substantially increased importance of the industrial sector has largely resulted from the growth of the oil sector rather than manufacturing. Indeed, the pattern of manufacturing development in South East Asia has, if anything, become more uneven (see chapter 5).

Agriculture's contribution to total production has, with the exception of Burma and Kampuchea, fallen throughout the region. However, for the pro-capitalist economies in general and Thailand in particular,

Table 1.14 *Changing structure of the South East Asian economies*

	Percentage share of GDP												Percentage share of the official labour force								
	Agriculture			Industry[1]			Manufacturing			Services			Agriculture			Industry[1]			Services		
	1965	1986	change	1965	1986	change	1965	1986	change	1965	1986	change	1965	1986	change	1965	1986	change	1965	1986	change
Burma	35	48	+13	13	13	0	9	10	+1	52	39	−13	64	53	−11	14	19	+5	23	28	+5
Indonesia	56	26	−30	13	32	+19	8	14	+6	31	42	+11	71	57	−14	9	13	+4	21	30	+9
Kampuchea	n/a	n/a	n/a	n/a	n/a	n/a	n/a	n/a	n/a	n/a	n/a	n/a	80	80	0	4	n/a	n/a	16	n/a	n/a
Laos	n/a	81	n/a	n/a	7	n/a	n/a	n/a	n/a	n/a	n/a	n/a	81	76	−5	5	7	+2	15	17	+2
Malaysia	28	20	−8	25	35	+10	9	21	+12	47	45	−2	59	42	−17	13	19	+6	29	39	+10
Philippines	26	26	0	28	32	+4	20	25	+5	46	42	−4	58	52	−6	16	16	0	26	33	+7
Singapore	3	1	−2	24	38	+14	15	27	+12	73	62	−11	6	2	−4	27	38	+11	68	61	−7
Thailand	35	17	−18	23	30	+7	14	21	+7	42	53	+11	82	71	−11	5	10	+5	13	19	+6
Vietnam	n/a	45	n/a	26	n/a	n/a	n/a	n/a	n/a	n/a	n/a	n/a	79	68	−11	6	12	+6	15	21	+6
Lower income	42	32	−10	28	35	+7	21	24	+3	30	32	+2	77	72	−5	9	13	+4	14	15	+1
Lower middle income	30	22	−8	25	30	+5	15	17	+2	43	46	+3	65	55	−10	12	16	+4	23	29	+6
Upper middle income	18	10	−8	37	40	+3	21	25	+4	46	50	+4	45	29	−16	23	31	+8	32	40	+8
Third World	30	19	−11	31	36	+5	20	22	+2	38	46	+8	70	62	−8	12	16	+4	18	22	+4

[1] including manufacturing
Source: World Bank, *World development report* (1988)

the proportion of the labour force directly engaged in agriculture remains high for the levels of per capita GDP (table 1.1 and 1.14).[18] The decline in the importance of the agricultural sector reflects the adoption of development strategies that have emphasised urban-industrial growth (see chapter 5) as well as the deterioration in the terms of trade discussed in 1.4.

A major consequence of the emphasis on urban-industrial growth and the relative neglect of the rural sector has been rapid urbanisation. While South East Asia (other than Singapore) has some of the lowest levels of urbanisation in the Third World the rates of growth (outside of the socialist group) are amongst the highest (table 1.15). Rural-urban migration adds to the still high rates of natural increase (table 1.16).

Since the late 1970s there have been sharp reductions in the rate of population growth in most South East Asian countries (table 1.16).[19] At the regional level, while the crude birth rate and crude death rate are only slightly below the average for the Third World, the infant mortality rate is substantially lower, a reflection of the generally better living conditions (table 1.1). There are, however, considerable variations within the region even outside of Singapore and the former Indo-China states (table 1.1). Malaysia, Thailand and the Philippines have similar crude death rates, but the birth rate decline has been most apparent in Thailand and least so in the Philippines. Malaysia's low infant mortality rate reflects the country's much higher per capita income and greater attention to rural development (see chapter 5). Burma and Indonesia have remarkably similar demographic characteristics with continued high birth, death and infant mortality rates.

The sharp falls in rates of population increase since the late 1970s (table 1.16) still leave Indonesia, Burma and the Philippines with the potential for rapid population growth. This prospect is most disturbing in Indonesia because of the sheer size of the population and its concentration in the Inner islands of Bali, Java and Madura (table 1.17). Elsewhere in the region pressure on land resources is less apparent. However, land shortages and serious environmental damage due to cultivation of unsuitable land are evident in many areas, including Thailand, the Outer Islands of Indonesia, the Philippines and Eastern Malaysia.

Much attention has focused on the concentration of South East Asian population in capital cities. This is most striking in the case of Thailand and, to a lesser extent, Burma, Laos, the Philippines and Vietnam (table 1.15). In contrast, Indonesia and Malaysia have well-developed urban hierarchies.

Allied to the rapid growth of urban areas has been the expansion of

Table 1.15 *Urbanisation*

	% of the population living in urban areas				annual average % rate of increase of the urban population			% of the urban population in the largest city	
	1960	1970	1980	1985	1960–70	1970–80	1980–85	1960	1980
Burma	19.3	22.8	23.5	24.0	4.0	4.2	2.8	23.0	23.0
Indonesia	14.6	17.1	22.1	25.0	3.8	4.9	2.3	20.0	23.0
Kampuchea	10.3	11.7	13.9	n/a	3.7	0.9	n/a	n/a	n/a
Laos	8.0	9.6	13.4	15.0	4.4	5.9	5.6	69.0	48.0
Malaysia	25.2	27.0	29.4	n/a	3.5	3.3	4.0	19.0	27.0
Philippines	30.2	33.0	37.4	39.0	3.8	3.8	3.2	27.0	30.0
Singapore	-	-	-	-	-	-	-	-	-
Thailand	12.5	13.2	17.0	18.0	3.6	3.3	3.2	65.0	69.0
Vietnam	14.7	18.3	19.3	20.0	4.4	2.9	3.4	32.0	21.0
Low income economies	17.0	n/a	21.0	22.0	4.2	4.4	4.0	10.0	16.0
Lower middle income economies	24.0	n/a	33.0	36.0	4.4	4.3	3.7	29.0	31.0
Upper middle income economies	45.0	n/a	63.0	65.0	4.2	3.8	3.2	27.0	26.0

Source: 1960, 1970 and 1980 United Nations figures cited by Jones, 1988: 137; 1985, World Bank, *World development report* 1988

Table 1.16 *Average annual growth rates of population by World Bank income group (percentage)*

	1960–70	1970–80	1980–86
Burma	2.2	2.2	2.0
Kampuchea	2.6	−0.6	3.1[1]
Laos	1.8	1.0	2.0
Vietnam	3.1	2.8	2.6
All low income	2.2	1.9	1.9
Indonesia	2.1	2.3	2.2
Philippines	3.0	2.7	2.5
Thailand	3.0	2.5	2.0
All lower-middle income	2.6	2.6	2.6
Malaysia	2.9	2.5	2.7
Singapore	2.4	1.5	1.1
All upper-middle income	2.5	2.2	1.9
China	2.3	1.5	1.2
India	2.3	2.0	2.2
All less-developed	2.4	2.1	2.0
South East Asia	2.8	2.7	2.1

Source: World Bank, *World development report* 1983 and 1988; United Nations, 1982; and various national census reports.
[1] United Nations estimate
In 1981 the Kampuchean growth rate was estimated to be 4.6–5.2 per cent, one of the highest in the world (Kiljunen, p. 34).

employment in the service sector. In the Philippines and Malaysia the increase has been accompanied by a fall in the service sector's share of GDP (table 1.14), a consequence of the low productivity of much of the expansion. The manufacturing sectors reflect the reverse situation, the increase in the sector's share of GDP exceeding that of employment. This is a result of the relatively high capital intensity of much of the region's recent industrial development (see chapter 5).

Behind the expansion of officially recorded manufacturing and service employment lies the largely unrecorded and often illicit informal sector. Despite attempts to suppress these activities in some major urban centres (see chapter 5) the informal sector remains a major provider of livelihoods, services and commodities (including housing) for the urban population. Indeed such 'petty commodity production' is an important underpinning of the low labour cost 'modern sector' (Santos, 1979).

Table 1.17 *Population distribution in Indonesia, 1983*

	Area (sq km)	% of total area	Population	% of total population	Density (per sq km)
Java and Madura	132,187	6.9	96,892,900	61.5	733.0
Sumatra	473,606	24.9	30,928,500	19.6	65.3
Kalimantan	539,460	28.3	7,350,000	4.7	13.6
Sulawesi (Celebes)	189,216	9.9	11,112,200	7.1	58.7
Bali			2,593,900	1.6	
Maluku (Moluccas)	570,100	29.9	1,534,300	0.9	19.7
Irian Java			1,268,600	0.8	
Others			5,814,600	3.7	
Indonesia	1,904,569	100	157,495,000	100	82.7

Source: Central Bureau of Statistics (1986).

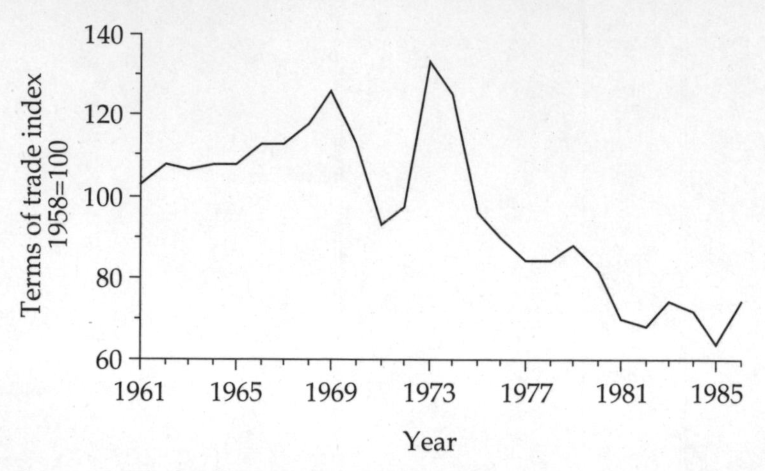

Figure 1.3 Thailand, terms of trade, 1961–86
Note: 1958 = 100
Source: Bank of Thailand, *Monthly Bulletin* Bangkok, various issues, *Changing terms of trade for Southeast Asia, 1978–86*

	1978	*1984*	*1986*
Burma	110	89	87
Indonesia	95	90	72
Malaysia	109	97	78
Philippines	98	58	52
Singapore	102	110	99
Thailand	87	52	52

Note: 1975 = 100
Source: Far Eastern Economic Review, *Year Book* Hongkong, 1983, 1987

1.6 The South East Asian economies and the international recession.

During the 1970s the world-economy entered a period of prolonged economic crisis. While there is not space here to discuss the nature, origins, or indeed, chronology of the crisis, it is important to stress that the crisis is truly global, affecting developed, less-developed and socialist countries (Thrift, 1986: 12). The timing and nature of the recession's impact have, however, shown considerable variation; to this the South East Asian states have been no exception. At the risk of over-simplification the impact of the recession on the South East Asian economies can be seen as a reflection of the nature of their interaction with the world capitalist economy (this theme is developed in chapter 5). Crises have been triggered (but not necessarily caused) by the changing price of oil, falling prices for South East Asian commodities, reductions in the

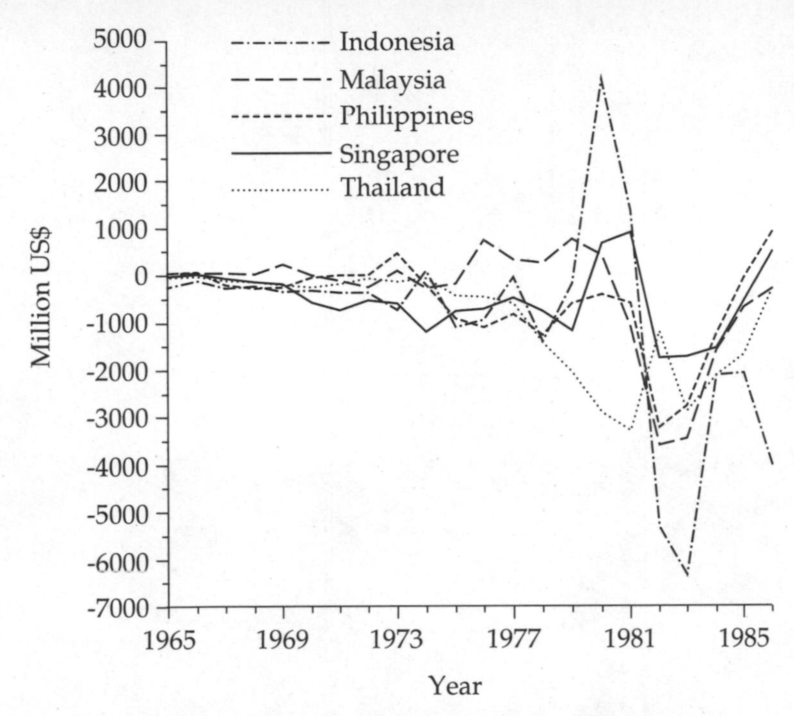

Figure 1.4 Balance of Payments on current account, 1965–86
Source: IMF, *International finance statistics year book Washington, various issues.*

demand for the region's manufactured goods and increasing cost and scarcity of foreign loans. In essence, as is argued in chapter 5, these changes brought deep-seated contradictions of the development process to the surface and accelerated such long-term trends as the deterioration of the terms of trade for the region's primary producers (figure 1.3).

From the early 1970s the fortunes of the South East Asian economies other than Singapore strongly reflected the movement of commodity prices (figure 1.2). The rise in oil prices in 1974 and 1979 brought crises to Thailand and the Philippines whilst boosting the fortunes of Indonesia and, to a lesser extent, Malaysia. The combination of the second oil price rise and the sharp fall in the price of other primary products brought Thailand and the Philippines to the verge of economic collapse, necessitating the acceptance of major interventions by the International Monetary Fund and World Bank. Similarly the reduction in oil prices since 1981 has eased the pressure on the importers but seriously undermined the economies of Malaysia and Indonesia and, to a degree, Indonesia.

Table 1.18 *Growth of foreign debt, 1970–86*

	1970		1975		1980		1986		% annual
	Total[1] debt as a % of GNP	Debt[2] Service Ratio	Total[1] debt as a % of GNP	Debt[2] Service Ratio	Total[1] debt as a % of GNP	Debt[2] Service Ratio	Total[1] debt as a % of GNP	Debt[2] Service Ratio	increase in debt 1980–86
Indonesia	37.7	7.8	27.4	7.5	22.5	8.0	33.4	8.3	21.1
Malaysia	12.9	2.6	14.9	3.3	13.6	2.3	47.8	6.3	52.4
Philippines	9.2	6.8	8.7	7.1	18.2	7.1	41.7	10.5	33.9
Singapore	9.3	0.7	9.9	0.7	13.2	1.1	10.1	2.4	12.1
Thailand	5.2	3.2	4.2	2.4	11.4	4.9	26.4	5.9	22.6
Brazil	7.7	14.8	11.2	16.8	16.6	34.5	35.5	21.5	22.6
Mexico	9.8	22.7	14.9	25.1	20.6	32.1	43.4	25.2	24.5
South Korea	24.8	19.2	27.4	11.3	28.2	12.3	35.0	6.5	14.3

[1] Debts Outstanding (DOD) as a percentage of GNP
[2] Total Debt Service (TDS) as a percentage of the value of exports of goods and services
Source: World Bank, *World Debt Tables*, Washington, various issues

Table 1.19 *Annual rate of growth of GDP, 1980–87*

	1980	1981	1982	1983	1984	1985	1986	1987
Indonesia	9.6	7.6	5.4	3.3	5.3	1.6	2.0	2.5
Malaysia	8.0	6.5	4.9	3.8	5.9	−1.5	0.0	3.0
Philippines	5.4	3.0	3.5	−1.4	−5.0	−4.0	1.0	5.7
Singapore	10.2	9.9	3.7	7.9	8.0	−2.9	2.0	8.8
Thailand	4.9	7.6	6.0	5.8	5.5	3.5	4.5	8.4
Hongkong	9.8	10.4	4.3	5.2	6.0	0.8	6.0	13.6
South Korea	−6.7	7.1	3.5	9.5	7.5	5.1	10.0	11.1
Taiwan	7.2	5.5	5.0	7.5	11.0	5.1	10.8	11.7

Source: These figures are based on a variety of national sources.

However, as was noted in 1.3, despite these problems the South East Asian states maintained comparatively high growth rates during the 1970s. In part this reflected the rather broader base of their economies compared to those of the Third World in general. Additionally, in the case of Thailand, a lower degree of integration with the world-economy provided a measure of insulation (table 1.10).

The maintenance of high rates of growth had a price, reflected in high levels of state expenditure (see chapter 5) and increased balance of payments deficits (figure 1.4). Since the region's credit rating remained high and overseas debt was of a low order these were increasingly met by foreign loans. To a degree growth in this period could be viewed as 'debt sustained'.[20] However, despite the rapid growth of foreign debt the debt-service ratios remained low by Third World standards (table 1.18) until the early 1980s.

During the 1980s economic growth has become more uncertain; even so, all the economies except the Philippines have achieved comparatively high rates of growth (table 1.19). In all except Thailand growth has at some point come to a shuddering halt – the Philippines in 1983, Burma, Indonesia, Malaysia and Singapore in 1985. Since 1981 there has been a very sharp increase in overseas debt (table 1.18), with debt-service ratios beginning to rival those of South Korea and Mexico.

The crises that have affected the South East Asian economies since the early 1970s have resulted in domestic and international pressure for major economic 're-structuring'. This appeared first and most clearly in Thailand where, following the crisis of 1981, the International Monetary Fund and World Bank imposed a programme of 'structural adjustment'. In essence this stabilised the economy while opening the country further to the activities of international capital. To a degree the South

East Asia states have been able to resist the pressure to restructure where this conflicts with the interests of domestic capital.

A combination of elements of restructuring, rethinking of national development strategy and a limited recovery of the world-economy has brought renewed high rates of growth to Thailand and Singapore. In Thailand, in particular, a new cycle of capitalist penetration and increased integration into the world-economy is occurring. However, the sustainability of this renewed growth is open to serious question.

2

Pre-colonial South East Asia

2.1 Introduction

Many Western accounts of South East Asian economic and social structures are both distorted and highly selective. The colonial powers fostered the view of backward, changeless societies with weak, ephemeral political structures, despotic, often rapacious, rulers, endemic disease, almost uniform poverty and internecine warfare.[1] This view remained implicit in much of the development literature of the 1950s and 1960s.[2] For advocates of 'modernisation theory' the traditional societies of the pre-colonial, colonial, and post-colonial periods were static, bounded and conservative. In the more extreme views such societies were presented as lacking economic logic (Rostow, 1960). Others, such as Foster (1965), accepted that traditional societies had their own logical basis. Similarly, Schultz (1964: 36–57) described traditional agricultural production as, within its own terms of reference, containing few 'allocative inefficiencies'.

For adherents to these views, pre-colonial South East Asia was dominated by communal, village-based production, over which weak centralised state structures had limited control and made limited demands. Production was dominated by agriculture to which craft activities were merely annexed. Such communities both aimed at, and largely achieved, self-sufficiency with little contact with the outside world (Murray, 1980: 6–7).

Over the last twenty years work by Western and Asian scholars has substantially undermined the picture of a static, undifferentiated, communal society. Increasingly it has been recognised that there was considerable variation within the region in the degree to which communal, individual and state structures had evolved. Insufficient attention is still paid to the processes of change. This is particularly true of the long period of Western contact that preceded colonisation. Between 1450 and 1800 there was substantial expansion of South East Asian trade and production (Evers, 1987; Reid, 1988). Trade became an increasingly important element in the process of state and core formation. Attention has tended to focus on the nature of the relations of production during this long period rather than on the processes of change that operated. However, there are 300 years of history during which South East Asia experienced considerable change as did the world economy.

2.2 South East Asian environments

South East Asia is a region of cultural and environmental diversity and complexity. In environmental terms a broad distinction may be made between 'Mainland' and 'Island' South East Asia. The mainland comprises a series of divergent mountain chains that enclose major depressions – the Mekong Valley, Central Plain of Thailand and the Irrawaddy Basin (map 2.1). These have formed the cores of major political units. None of the mountain chains are of great elevation; only the Arakan Yoma (which exceeds 3,000 metres) appears to have offered a serious barrier to communication. Spencer (1973, 28–31), however, suggested the barriers were sufficient to allow early political structures to become established within the basins but posed no obstacle to their becoming the cores of early imperial structures that could dominate much of the mainland and, on occasion, hold sway over parts of Island South East Asia. The earliest of these was Funan, based in present-day Kampuchea which, at its height at the end of the fifth century, extended its control into southern and central Vietnam, central and southern Thailand and over much of the Malay Peninsular. The core of this state, the Tonlé Sap basin, subsequently became the focus for the Khmer empire with its capital at Angkor. At its height in the twelfth century Khmer control extended over much of present-day Thailand, Laos, Kampuchea and parts of Vietnam and Burma.

Island South East Asia is structurally much more complex (map 2.1).[3] The small size of river valleys in the islands and the restricted areas of alluvial lowlands outside of eastern Sumatra and Kalimantan limited the agricultural resource base available to early states. Thus, Island South East Asia has shown a much more fluid and fragmentary pattern than the mainland (Spencer 1973: 30–5). Where large political structures did emerge, their power rested on trade and control of the seaways (Fisher, 1964: 106). This development was seen at its most spectacular between the seventh and thirteenth centuries in the rise of Srivijaya with

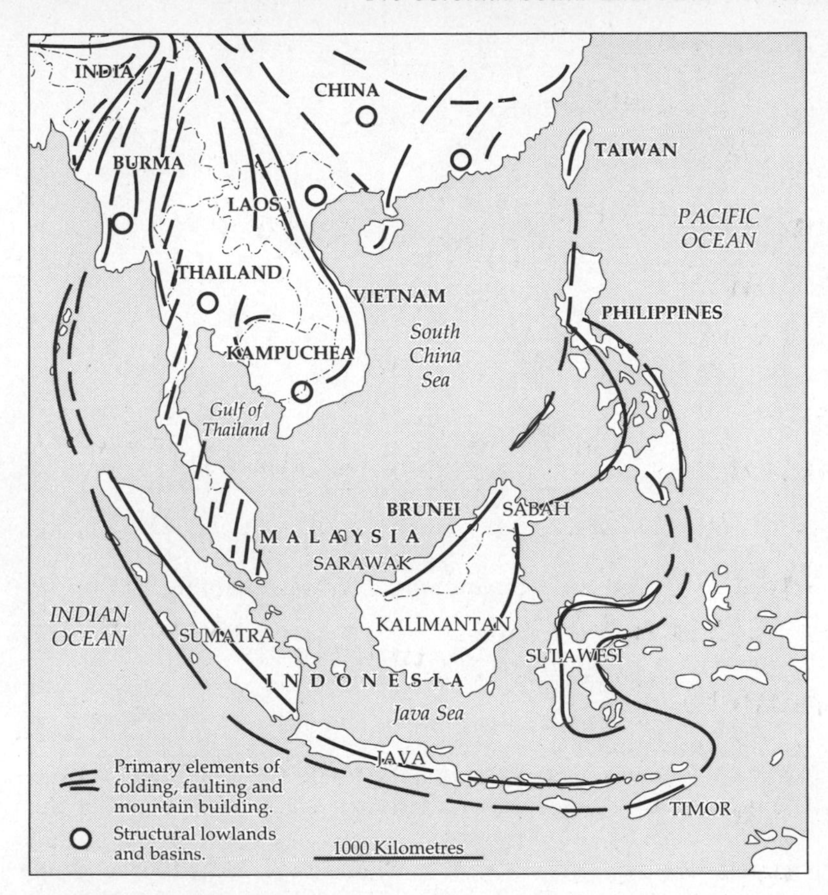

Map 2.1 Structure of South East Asia
Source: Spencer, 1973: 5–6

its river-port capital at Palembang in southern Sumatra.[4] The commanding position on the sea route between India and China made it the first in a succession of port-cities – Malacca (Melaka), Aceh, Penang and Singapore, whose strength was based on the control of the Straits of Malacca (Harrison, 1963: 22–3). At its height Srivijaya controlled not only much of southern Sumatra, but the hundreds of islands that dot the approaches to the Straits of Malacca and land on both sides of the straits (Andaya and Andaya, 1982: 19). Only in east-central Java did a cluster of small, exceptionally fertile river basins enable a sufficiently large and regular surplus to be produced for states to form along the lines of the mainland pattern (van Leur, 1955: 105).

In climatic terms it is rainfall that has the greatest influence on the

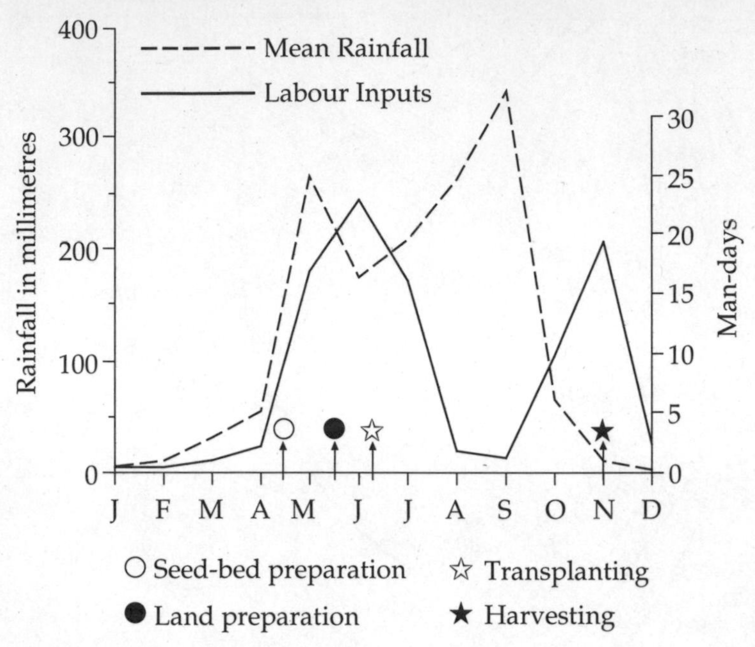

Figure 2.1 Seasonal activity in Thai rice cultivation
Source: Dixon, 1990a: 24.

pattern of human activity – the minimal temperature variations impose few constraints on agricultural activities.[5] While the majority of the region receives over 1500mm of rain annually, seasonal and spatial variations are much greater than for temperature. Indeed for much of the region the seasonal distribution of rainfall exerts a decisive influence on human activity (see figure 2.1).

2.3 Food and agriculture

South East Asian agriculture is dominated by the cultivation of wet rice. Compared to similar climatic areas of Africa and the Americas the variation in crop complexes is much less. This cannot be merely explained in terms of environmental conditions. The physical configuration and location of the region facilitated widespread cultural contact. Indeed despite the high average rainfall levels for much of the region variations in annual totals and more especially the seasonal distribution poses problems for wet-rice cultivation. There are 'few parts of the region where rainfall is either heavy or reliable enough to ensure a good crop in every season' (Fryer, 1979: 49). In the northern

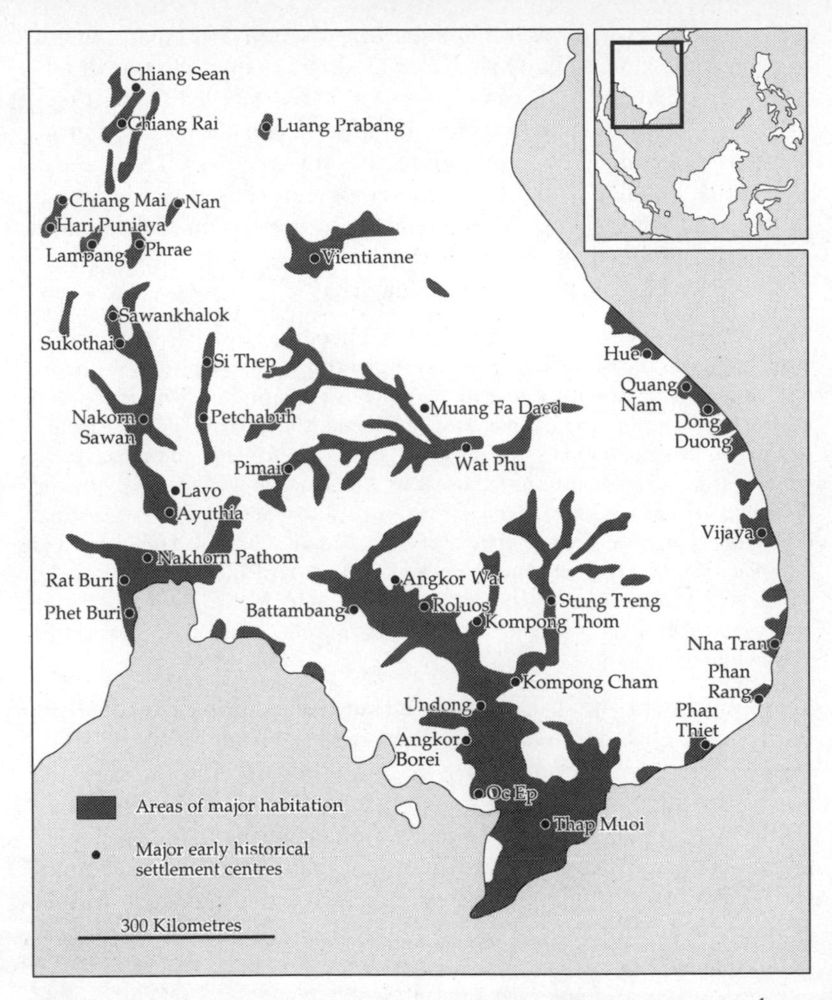

Map 2.2 Major early historical settlement centres and congenial habitats[1]
Notes: [1]based on suitability for paddy cultivation and availability of potable water
Source: Ng, R. C. Y. 1979: 271

fringes of the region, lower and less reliable rainfall results in the dominant wet-rice cultivation being highly marginal.

Rice cultivation has been long established in the region[6] and formed the basis of all the major pre-colonial mainland cultures. Map 2.2 illustrates the general relationship between the areas most suited for wet-rice cultivation and early settlement cores.

The earliest cultural niches occupied by wet-rice cultivators were not the potentially most productive alluvial areas of the major river basins.

Early cultures lacked the technology, organisation and labour supply to exploit these areas. Early sites were in drier, less fertile areas of reliable rainfall where land was easily cleared and levelled (Ishii, 1978). The cores of early cultures gradually moved into the more environmentally demanding, but more productive, lower valleys. These moves were made possible by improved technology, organisation and labour supply, and made necessary by increases in population and the need for surplus production.

As Ng R.C.Y. (1979: 270–1) has noted:

> The most extensive of these habitats, the Cambodian Lowlands, provides the stage for much of the dominant stream of South East Asian history – the rise and fall, in succession, of several of the most important civilizations and centres of political power, such as Funan, Chenla, Champa and finally Angkor. It can also be seen that once the migrant Thai began to settle and organise the western tract, extending from Sukothai to Ayuthia, the way to the south and their potential as a dominant power was opened for them . . . the earlier kingdoms of Nan and Kampengphet, of Chiangmai, Phayao, Lumphun and Lampang, faded away eventually under the impact of the southern powers, which had finally organised their extensive habitats for irrigated agriculture and were capable of producing an increasing surplus of grain for supporting armies and artisans.

In contrast to the mainland, millet and tubers long remained the staples of the islands and much of the Malay Peninsula (Bray, 1986: 10–11). Irrigated rice may have been introduced into Java in the twelfth century. From there it spread through the Malaysian Peninsula (Hill, 1977: 20–27). The dry cropping systems of the islands probably had a productivity of little more than half that of even the most unintensive wet-rice system.[7] This almost certainly mitigated further against the development of major political units based on large agricultural surpluses.

Even the most simple attempts at water control[8] necessitate some form of communal organisation. The basis for the major mainland political units which developed in the pre-colonial period was extensive organised water control (Fisher, 1964: 93–4; Ng R.C.Y., 1979: 265–66).[9] The Khmer Empire was dependent on the large and regular production of surplus rice in the Tònlé Sap.[10] This area, although one of the most inherently fertile in South East Asia, is only exposed for cultivation during a short period of the dry season. Thus the essence of the Angkor system was the large-scale storage of water for regulated release during the planting season (Ng R.C.Y., 1979: 265). This enabled not only very productive single, but also double, cropping to take place (Groslier, 1974: 103). Indeed multiple cropping appears to have been established in the pre-Khmer period in Champa (southern Vietnam) as early as the first century AD (Bray, 1986: 292).

Similar arguments can be advanced for the Ayuthia period in Thailand, the Annamite Kingdom in the Red River Delta and the Burmese culture in the central dry belt. Many of these systems were not only extremely large but very long-lived; they were extended and elaborated by successive cultures. Much of the extensive canal system on which the Khmer Empire depended was built by Funan (Bray, 1986: 73). According to Groslier (1979: 190), the Angkor system at its height irrigated 167,000 hectares along the northern side of the Tonlé Sap alone. The individual tanks in the Burmese Dry Belt, constructed between the eighth and the thirteenth centuries, irrigated as much as 15,000 hectares (Stargardt, 1983: 1945).

The development of water control was of major importance in the long-term increase in the productivity of wet-rice cultivation. However, the productivity of rice can be increased substantially by low-cost, often inconspicuous, improvements. Gradual improvements in land levelling and water control as a result of long-term cultivation and perhaps increased labour input can achieve significant increases in productivity. More striking are the results of replacing broadcasting, transplanting, double ploughing and harrowing. Wet-rice cultivation must be regarded as a highly dynamic system under which a large volume of additional labour can be absorbed before diminishing returns set in. The ability to do this without any major technical change has blinded many to this dynamism, labelling pre-capitalist South East Asian production systems as static if not stagnant.

Not only were increasingly large numbers of agricultural households supported in the core areas but reliable surplus production expanded to provide for major urban centres, large-scale construction, non-agricultural production and extensive trading networks. Indeed the introduction of irrigation not only increased productivity but, perhaps more significantly, made production more reliable. Reliability is crucial in the development of urban centres and major state apparatus. Many of these systems supported population densities substantially higher than those that prevail today (Delvert, 1961; Stargardt, 1983: 36–9).

It is tempting to relate the changing intensity of rice cultivation to population density. In areas of abundant land there was little incentive to intensify production. Intensification for most of the region necessitated restricted access to land and organisation above that of the village community to establish the necessary water control. Only in small basins and tributary valleys, of for example northern Thailand, was village-level organisation sufficient to establish the necessary water control. It is, however, important not to become preoccupied with water control as a key to understanding the region's societies. As Bray (1986: XV) has stressed:

> Of course the presence of irrigation works does affect technical and social organisation ... But ... the presence *per se* of irrigation works is not sufficient to explain many of the features of Asian societies. More significant is the association of irrigation with labour-intensive methods of cultivation and, more particularly, with the labour-intensive cultivation of rice.

The nature of pre-colonial South East Asia production systems is examined in 2.5 and 2.6.

2.4 Population and culture

The region's character as a focus of converging land and sea routes has brought long-term movements of people overland from the north, north-east and north-west, together with coastal and sea movements from both the east and west.

In the pre-colonial period the peopling of South East Asia took place through a series of long-term southerly migrations. In the main these had their origin in the continental interior, percolating into mainland South East Asia and then 'island hopping' into present-day Indonesia. These movements were probably supplemented by migrations along the off-shore island arcs. Some of the early movements continued beyond the region's limits into the Pacific. Successive waves of migrants have occupied the favoured 'ecological niches' in the coastal and riverine alluvial lowlands, displacing many of the existing people, who moved either into the upland areas or further south into the outer islands. However, considerable intermixture has taken place with new migrants being absorbed by the older, and in the long term perhaps only adding linguistic, technical and cultural elements to those of the established population. The situation in the mainland is far more complex, a result of major movements until the twentieth century and minor ones until the present (map 2.3 and table 2.1).

The mountain barrier that separates Assam from mainland South East Asia has limited overland contact with India. But as early as 600 BC the traders and explorers of the Coromandel coast were in contact with Burma, Peninsular Malaysia and the western islands of Indonesia (Wheatley, 1955: 65–67). In the early centuries AD increased trade between South East Asia and Southern India was accompanied by the setting up of small colonies of merchants at key points on the sea trade routes. Hall (1981) discusses conflicting views of the process

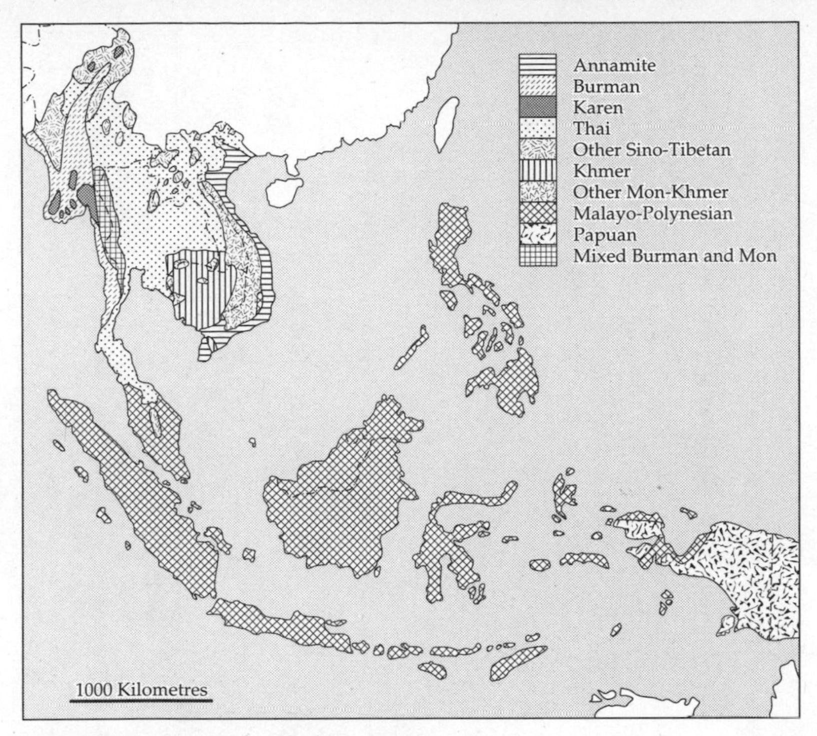

Map 2.3 Languages of indigenous peoples
Source: Dixon, 1990b, based on Fisher (1964: 99)

which led from trading to large-scale cultural influence in mainland South East Asia and Indonesia. 'Indianized' states, such as Champa and Funan, existed on the mainland by the second century AD, and Indian cultural influences spread widely until AD 1000.[11] Concepts of monarchy, social order and urban planning as well as religion (particularly Buddhism), literature and script, all of Indian origin, spread along the sea routes into most of mainland South East Asia and into eastern Indonesia. The eastern islands and the Philippines received these influences indirectly. The present-day states of Burma, Thailand, Laos and Kampuchea owe much to this period of Indian cultural influence.[12] In contrast the pre-colonial influence of China was largely limited to direct overland contact. Chinese conquest of north Annam in 111 BC initiated a period of control that lasted until AD 939 and which stamped a lasting impression on the landscape and culture (Fitzgerald, 1972: 19–38). Chinese methods of wet-rice cultivation, water control, dyke-building and the use of fertilizer of human, animal and vegetable origin, were all introduced. The Annam region stood out from other wet-rice

Table 2.1 *Estimates of ethnic composition*[1]

Percentage	> 80	> 60	> 40	> 20	> 10	> 5	> 1
Brunei			Malay	Chinese		Kadazan	
Burma		Burmese				Koren	Chinese
						Shan[2]	Kachins
							Other hill people
Kampuchea	Khmer					Vietnamese	Chems
							Chinese
Indonesia		Javanese			Sudanese	Medureso	Chinese
							Balak
							Bubinese
							Dyaks
Laos			Lao[2]				
			Montagnard[3]				
Malaysia			Malay	Chinese		Indian	
Philippines				Cabuano[4]	Iloko	Bikol	
				Tagalog	Panay-Hiligaynon	Bisaya	
Singapore		Chinese			Malay	Indian	
Thailand	Thai[2]				Chinese		Hill people[5]
Vietnam	Vietnamese					Lao[2]	
						Montagnard[3]	
						Chinese	

1. These proportions are based on a variety of estimates.
2. Thai-Lao-Shan belong to the same ethnic group.
3. Montagnard is a French collective term for hill peoples, in particular the Muong, Meo, Yao and Thai.
4. Cabuano is the first language of 24 per cent. Tagalog of about 21 per cent.

Source: Based on Neville (1979).

5. The Thai Hill people population is probably less than half a million, over 50 per cent of which are Karen and 20–25 per cent Mao. Present in small numbers are the Yao, Lisu, Lahu, Akha, Lawa and Khamu.
6. Including Chin, Lolo, Naga and Pajaung Wa.
7. There are in total 20 minor groups with estimated populations of 1–2 million.

areas of South East Asia in having a more sophisticated and durable production system. From the late nineteenth century the French colonial presence to a degree 'de-sinified' Vietnam, for example, by romanising the script.

Elsewhere in South East Asia the influence of early Chinese culture is much less evident. However, important cultural traits of Sinitic origin were introduced into mainland South East Asia by the migrations of the Shans, Laos, Burmese and Thais who settled for long periods in areas under some degree of Chinese control before expansion and population growth drove them south (Fisher, 1971: 89). The Chinese culture brought to the region was to a degree overlain by the heavily Indianised cultures of the early migrants, such as the Mons, Pyus and Khmers. However, many elements of Chinese culture persisted, as in methods of water control for wet-rice. In contrast, it is usually asserted that the islands of South East Asia felt little Chinese influence (Purcell, 1965: 17). However, it may well be that the preoccupation with 'cultural traits' has resulted in a serious underestimating of the influence of the Chinese in the islands and, indeed, the region as a whole (Fitzgerald, 1972: 135; Salmon, 1981). Chinese trade, exploitation of resources and introduction of technology made a major contribution to the region's economies long before the nineteenth century.

The lasting pre-colonial cultural influence in the islands of South East Asia and the Malaysian Peninsula was neither Chinese nor Indian but Arab. Muslim merchants appear to have frequented South East Asian ports from at least the seventh century AD, but it was not until the thirteenth century that Islam began to be adopted in Sumatra, spreading along the Straits of Malacca with increasingly dominant Gujerati and Bengali merchants. As Harrison (1963: 51–2) has succinctly stated:

> the acceptance of Islam by the local rulers of those coastal areas must have been largely inspired by the wealth, commercial success and good business sense which those merchants represented.

During the fourteenth and fifteenth centuries, Islam spread rapidly into Java and the Malaysian Peninsula (Harrison, 1963: 55–60), thereby producing the basis for the present-day dominance of the religion in Indonesia and Malaysia.

South East Asia was a region of remarkable physical, cultural and ethnic complexity before Western colonisation. The political and economic structures are usually presented as having been in a state of constant flux. While this was more true in the islands than on the mainland, it is difficult to argue that Europe, even in the modern period, has been significantly more stable in this respect. 'Fluid political bound-

Map 2.4 South East Asia: political divisions in the late fifteenth century
Source: Fisher, 1964: 118; Reid, 1988: 9

aries and endemic warfare' is a description that could equally well apply to eighteenth-century Europe. Map 2.4 provides a general indication of the major political structures in the region on the eve of the penetration of the European powers.

2.5 The organisation of production

A distinction can be made between the major political units and the areas peripheral to them where little organisation above the village level had developed. As was noted in 2.2, environmental conditions, location with respect to trade routes, population density, ethnic composition and crop complexes were of considerable importance in this distinction. However, the persistence of the peripheral societies also has to be understood in terms of their relationships with the major states. The ebb and flow of political power, particularly amongst the mainland states, brought the peoples of the bordering area under varying degrees of control. In certain periods considerable surplus produce and labour was extracted from these communities. While the pace of change in these societies was slow they cannot be regarded as static. New crops, techniques and organisation of labour were adopted to increase production to meet the demands of population growth, loss of land to major lowland states, trade and external levies.

Most attention has been directed at the organisation of production in the major political units. Here most of the debate has focused on the applicability of particular formulations. While most writers have concluded that these societies exhibited feudal characteristics there has been considerable debate over the extent to which an abstraction from a comparatively short period of European history can be meaningfully applied to other parts of the world (Wolf, 1982: 81). This has led to renewed interest in the utility of Marx's formulation of the AMP (Asiatic Mode of Production; see Bailey and Llobera, 1981a; Krader, 1975; Melotti, 1977).

Wittfogel's (1957) study of 'oriental despotism' and 'hydraulic society' was a major factor in the renewal of interest in the Asiatic Mode of Production. Behind Wittfogel's exposition lay 'a form of geographical or ecological determinism, in which the forces of production determining a society's social and political organisation are themselves determined by climate, soil, hydrography, etc.' (Bailey and Llobera, 1981c: 109). Wittfogel suggested that in a range of Asian countries where agriculture was based on large-scale irrigation which necessitated centralised control ('agro-managerial') a despotic state apparatus emerged. While these views have been heavily critisised on both theoretical and empirical grounds, they have been a major influence on the debate over pre-capitalist social formations (Eberhard, 1958; Leach, 1959; Steward, 1977).

Debate has surrounded the nature, applicability and utility of the Asiatic Mode of Production. To a degree this has stemmed from the partial and generally fragmentary development of the concept in the writings of Marx and Engels (Bailey and Llobera, 1981b: 23–52). The original formulation of the AMP was of a social formation that, unlike European feudalism, was incapable of making the transition to capitalism. That is, the AMP was a kind of 'blind alley' of development which led to stagnation. More recent adherents have extended the applicability of the concept to much of Africa and South America (Varga, 1968; Godelier, 1978).

The main features of the AMP have been summarised by Mandel (1971: 121–122):[13]

(1) What is above all characteristic of the Asiatic mode of production is the absence of private ownership of land.
(2) As a result, the village community retains an essential cohesive force which has withstood the bloodiest of conquests through the ages.
(3) This internal cohesion of the ancient village community is further increased by the close union of agriculture and craft industry that exists in it.
(4) For geographic and climatic reasons, however, the prosperity of agriculture in these regions requires impressive hydraulic works . . .
(5) For this reason, the state succeeds in concentrating the greater part of the social surplus product in its own hands, which causes the appearance of social strata maintained by this surplus . . . [There exists] a very great degree of stability in basic production relations.

The number of studies that have attempted to fit the AMP to specific pre-capitalist states in South East Asia are few. Elliott (1978) made an interesting attempt to apply the AMP to the analysis of pre-capitalist Thailand. There is in the study a general assumption of the 'goodness of fit' of the formulation, though the author does not make any detailed appraisal of this. He does, however, comment on the 'contamination' of the mode of production by the continued existence of slavery and the possibility of a 'feudalistic class' emerging from the stratum of state functionaries (Elliott, 1978: 39). In addition, other studies (see 2.7) suggest that private ownership of land was well established in Thailand by the early nineteenth century.

In contrast Khoi (1973) examined the applicability of the AMP to Vietnam and found that:

> The example of ancient Vietnam is all the more interesting since, far from confirming the use of the concept of the 'Asiatic mode of production', it contradicts certain characteristics generally attributed to the notion, particularly the absence of private property and even (for some authors) the absence of social classes. On the other hand, Vietnam does not reveal the classic features of 'feudalism' either; over the period from the tenth to the nineteenth century there is no observable evolutionary tendency toward 'feudalism' in this sense. On the contrary, the inverse is true, due largely to the strengthening of the literati-official class.

Additional weight is given to Khoi's conclusions that private ownership of land was present and indeed increasing in the early nineteenth century by Murray (1980: 50).

There is, then, little evidence that the AMP reflects the situation in pre-colonial South East Asia. As is discussed in 2.7, evidence points to the widespread establishment of ownership of land and indeed many 'capitalist features'. It is clear that many of these features do not fit neatly into either the Feudal or Asiatic modes.

The importance of the mode of production in understanding pre-capitalist societies lies not in attempts to test 'goodness of fit' but with Marx's method. As Chandra (1981) has advocated, attention should be devoted to the AMP and Marx's writings on Asia merely to inform ourselves of the method of analysis employed. It is on this basis that a number of writers (Amin, 1973; Wolf, 1982) have advocated the adoption of a broader formulation in which the Feudal and Asiatic modes may be regarded as polar cases. Under this, the South East Asian societies may be regarded as variations on situations where labour was mobilised and surplus extracted by the exercise of power and domination rather than economic pressure (Marx, 1859: 790–1; Wolf, 1982: 80; Vasiliev and Stuchevski, 1967; Töpfer, 1974), Amin (1973) has referred to this as the 'tributary mode of production' within which the principal variations are the degree of centralisation of surplus extraction and the role of mercantile trade.

2.6 Community and state

The importance that most writers have attached to communal organisations in pre-colonial South East Asia is illustrated by Clifford Geertz's (1963) study of Javanese peasant society. Geertz placed great emphasis on the communal nature of pre-colonial Javanese life. This study exerted a decisive influence over views of not only pre-colonial Java but of South East Asia as a whole. A number of more recent studies have questioned this view, most notably Onghokham (1975) who argued that the dominance of rural Java by communal village-based production was itself a product of the Dutch colonial system.[14] The development of individual household operation of land was reversed and communal ownership reinforced. This is not to deny the importance of communal, 'clan' and extended family land-holding and production (Oki, 1984: 278–9). Indeed both wet-rice and shifting systems of cultivation frequently necessitated labour inputs beyond the capacity of individual households (Murray, 1980: 52). However, it is clear that communal organisations were neither static nor the only ones in operation in the region.

The mechanisms that operated within the village communal system and its relationship with the state in South East Asia is more fully examined by Scott (1976), principally in the context of pre-colonial French Indo-China and Burma. For Scott, the key element is the fear of food shortage which has given rise to the 'subsistence ethic':

This ethic, which South East Asian peasants shared with their counterparts in nineteenth-century France, Russia, and Italy, was a consequence of living so close to the margin. A bad crop would not only mean short rations; the price of eating might be the humiliation of an onerous dependence or the sale of some land or livestock which reduced the odds of achieving an adequate subsistence the following year. The peasant family's problem, put starkly, was to produce enough rice to feed the household, buy a few necessities such as salt and cloth, and meet the irreducible claims of outsiders. The amount of rice a family could produce was partly in the hands of fate, but the local tradition of seed varieties, planting techniques, and timing was designed over centuries of trial and error to produce the most stable and reliable yield possible under the circumstances. These were the technical arrangements evolved by the peasantry to iron out the 'ripples that might drown a man'.[15] Many social arrangements served the same purpose. Patterns of reciprocity, forced generosity, communal land, and work-sharing helped to even out the inevitable troughs in a family's resources which might otherwise have thrown them below subsistence. (Scott, 1976: 2–3)

Thus behind a variety of technical, social and economic arrangements there is a strong element of insurance (Lipton, 1968: 341). Communities are, in effect, conditioned by what Scott (1976: 5) calls the 'safety first principle'. The operation of this is most clearly seen in those areas of South East Asia which experience the greatest environmental uncertainty.

While the 'safety first principle' can be shown to govern a wide range of activities, including trade and manufacturing, its operation does not imply that such societies were static. Indeed, South East Asian peasant communities have shown themselves receptive to a wide range of new crops, techniques and means of livelihood.[16] While a great many developments were brought into the region, particularly from India and China, this should not blind us to the capacity of the region's pre-capitalist systems of production to innovate.

The stability of the pre-capitalist state rested on the regular extraction of surplus from the peasantry and the insurance of the subsistence minimum (Wolf, 1969: 279). While generalisation from the limited studies available has to be approached with caution it seems likely that external claims on the community were proportional, flexible and slight compared to later colonial exploitation (Wales, 1934; Scott, 1976). Evidence points to the demands of the pre-colonial state being gauged to maintain control over the surplus-producing peasantry. Failure to ensure the subsistence minimum resulted in unrest, rebellion and, in a situation of relative abundance of land, migration out of the state's sphere of control. In some periods of crisis the state or members of the ruling class provided relief to enable the surplus-producing peasantry to survive (Somersaid Moertono, 1968). Surplus was extracted from the peasantry to support the ruling class, provide for public works, administration, warfare, and to supply goods for overseas trade. Levies took the form of labour services and/or payments in cash or kind.

The scale of the surplus extracted is impossible to gauge with any certainty.[17] However, it was sufficient to support the court and bureaucracy, maintain major urban centres, fund substantial public works, regular international trade and, at times, major military campaigns.

By 1850 Bangkok probably had a population of over 100,000[18] (Sternstein, 1984), a remarkable figure given that the total population of Thailand was only 4.5–5 million. Sufficient surplus rice was produced to support the capital and to provide for exports – in favourable years perhaps as much as 5 per cent of the crop (Ingram, 1971: 29). Certainly rice was being regularly traded with China and Java during the eighteenth century (Sarasin Viraphol, 1977: 70–138). The main public works had been the construction of canals for irrigation and transport. Much of the present extensive system was built before 1850 (Takaya, 1987: 182–90). The productivity and relative reliability of the rice-growing areas of the Central Plain that so impressed observers (Crawfurd, 1828: 420–1), was very largely due to these extensive works. The canal network made not only the production of the surplus possible but also enabled it to be transported for urban consumption and export.

While South East Asian pre-colonial communities may have been dominated by a 'subsistence ethic' this does not imply that they were self-sufficient. Most communities engaged in some form of exchange to obtain goods that they lacked. Western Sumatra,[19] for example, prior to Dutch annexation in 1830, was exporting rice, gold, pepper, coffee and camphor in exchange for raw cotton, cotton cloth, earthenware and iron bars. As well as a range of handicrafts for subsistence and limited exchange, more specialised manufacture had developed. Most notably, textiles, mats, paper, agricultural implements and firearms. A high degree of village specialisation in some of these products had developed with trade links that extended throughout Western Sumatra and beyond using the ports of the western coast and river routes east to the Straits of Malacca.

The West Sumatran situation was probably not untypical of much of the region on the eve of Western annexation. However, there were also areas where a much greater degree of specialisation was of long-standing. These activities included mining, metal-working, pottery, salt manufacture, textiles, shipbuilding, production of specialised crops and collection of a variety of natural products, all producing commodities that entered national, regional and international trading networks. In some instances specialisation was such that the community was dependent on external supplies of basic items. This is most notable in the production of spices – cloves, pepper, nutmeg and mace – in the Celebes (Sulawesi) and Moluccas (Maluku). Many islands were dependent on food imports, particularly rice from Java. This gave the Javanese entry into the islands and provided the basis of their political power (Tate, 1971: 51).

Textiles from outside the region were important items of trade well before the penetration of Western traders. In some places, such as the islands of Banda, imported goods displaced local products (Owen, 1978: 158). The early European traders found that Indian cotton goods were a common medium of exchange throughout the region, particularly for spices (Meilink-Roelofsz, 1962: 94–6, 161 and 214). By the seventeenth century most Javanese batik was either woven from imported yarn or merely dyed Indian muslin (Furnivall, 1936: 373–4). Thus when European mass-produced textiles began to reach South East Asia during the nineteenth century, they at first replaced Indian imports rather than local products (Owen, 1978: 158).

These systems of production and exchange produced a living standard that for the mass of the people was at least the equal of that prevailing in Western Europe. Reid (1988: 48–9) has concluded that the fragmentary evidence points to South East Asian people being of similar stature to Europeans with greater life expectations and perhaps lower infant mortality rates:

> The relatively good health of South East Asians in the age of commerce [1450–1680] should not surprise us if we compare their diet, medicine, and hygiene with those of contemporary Europeans. For the great majority of South East Asians serious hunger or malnutrition was never a danger. The basic daily adult requirement of one kati (625 grams) of rice a day was not difficult to produce in the country or to buy in the city, and it contained in itself enough calories and protein for healthy development. The relative lack of animal proteins was probably on balance an advantage, as it spared South East Asians from diseases spread through maggoty meat. Large-scale famine appears to have occurred only as a result of warfare.

During the nineteenth century the development of European nutrition, medicine and hygiene on the one hand and the imposition of colonial rule on the other drastically altered the balance. While it is dangerous to generalise, Ng Shui Meng's (1979: 56–7) study of one district in Luzon revealed that life expectancy fell from 42 between 1805 and 1820 to 17.5 in 1900 (see 4.4).

2.7 Changing systems of production

The process of change in pre-colonial South East Asia is probably best documented in the case of Thailand. Here a number of studies suggested that while the *form* of many pre-capitalist institutions and practices showed considerable continuity their *content* underwent profound change.

During the first half of the nineteenth century, in theory the ownership of Thailand's land and people continued to be vested in the King. However, in practice, private ownership of land and, for many, freedom from labour services, had emerged well before the involvement of the Western powers. The right for a freeman to take as much uncultivated land as he and his family could operate (usually no more than 4 hectares) was firmly established. Various regulations and procedures had been established which involved the would-be cultivator in a cash outlay.[20] Once these were fulfilled, while in theory the right to the land remained usufructuary, in practice the cultivator had full rights and land might be sold or mortgaged (Ingram, 1971: 12).

Theoretically all people were chattels of the king and freemen were obliged to perform personal services as well as to turn over part of their produce (Chatthip Nartsupha and Suthy Prasartset, 1981: 27). In practice, by the nineteenth century much of this surplus accrued to other members of the ruling class who acted as 'patrons'.[21] These in turn were obliged to the Crown for cash, kind and labour demands, but rather less than they extracted from the peasantry (Ingram, 1971: 13). By the nineteenth century obligation to the Crown had been widely commuted into cash payments (Hong Lysa, 1984: 52).

The nature of the patron system has been summarised by Ingram (1971: 13):

> The relation between the patron and freeman was personal, not territorial. This was an essential part of the feudal system of Siam after the fifteenth century. Every freeman had to have a patron to whom he was obligated for produce and services, but the freeman could choose his own patron and could move from place to place. In return, the patron had the duty of protecting his clients (as in matters of justice) and of lending them money when they needed it. If he did not do the latter, they might sell themselves into slavery, in which case the patron lost their services completely.

As a result of royal land grants, purchase or acquisition of uncultivated land, large holdings were accumulated by members of the ruling class. These were cultivated using slaves and labour dues from freemen (Ingram, 1971: 14). Given the relative abundance and ease of access to land there was little development of tenancy in Thailand before 1850 (van der Heide, 1906: 3–4). Control over labour was the key element in the pre-capitalist Thai system of production.

> In the pre-modern, agrarian society that was Siam in the early nineteenth century, the scarcity of manpower was a factor that had to be taken into account by the king in almost all the military, political, and economic calculations that he made. The shortage of labour to cultivate the land, to fight off enemy invasions, and to make up the work-force for construction projects made people a precious resource which the king tried to increase by foreign conquest, enticement, and compulsion. Equally as important as the promotion of settlement of population in the country was the ability to mobilize the people and the products of their labour for the state's use. For this purpose, the whole framework of social organization was centred on the control of manpower. (Hong Lysa, 1984: 9)

The change in the basis of the Thai social formation away from control over land and the associated commutation of obligations had also taken place in Vietnam by the early nineteenth century (Murray, 1980: 49–53). The two states differed in the details of the social relations of production, for example in the development of cash rents for land in Vietnam.[22] Similarly the Vietnamese administrative structure, modelled as it was on the Chinese system, differed markedly from that of Thailand. However, the reduction of the territorial basis of power in both states concentrated power in the hands of the Crown, limiting the power of the remainder of the ruling class to challenge that authority and increasing the long-term stability of the state. It would be dangerous to infer that these developments were common to the region as a whole. However, Warren (1985: 131) concludes that:

> There is general consensus that access to land was not a significant problem throughout South East Asia until the nineteenth century and that control over labour was most important in defining the nature of pre-capitalist relations of production.

In general attention has focused on the development of more productive and stable agricultural systems, the scale and manner of extraction of surplus and the mechanisms whereby it was redistributed. However, the process of change cannot be meaningfully understood by studying these societies in isolation. It is now appreciated that the expansion of regional and international trade played a major role in the changing organisation of production and the formation of state power (Evers, 1987). While most studies (Braudel, 1979; Reid, 1988 and Wolf, 1982) have stressed the importance of trade from the fifteenth century, others, notably Abu-Lughod (1988) have drawn attention to its earlier significance.

It may well be that commercialisation and monetarisation of the economy had spread much more widely in early-nineteenth-century South East Asia than most authorities have suggested. In Thailand production for the international market appears to have expanded rapidly from the late eighteenth century (Hong Lysa, 1984: 38–74; Nidhi Aeusrivongse, 1982: 75; Evers, 1987: 762–3; Sarasin Viraphol, 1977: 177). Indeed Nidhi Aeusrivongse (1982: 75) goes so far as to suggest that 'the production of agricultural goods for export can be regarded as the basis of the economy. As such the Thai economy can be defined as an exchange economy long before the time of the Bowring Treaty [1855].' A similar argument is advanced by Evers (1987).

Most studies of Thailand in the early nineteenth century emphasise the persistence of state monopolies, particularly in international trade, as barriers to modernisation. However, Hong Lysa (1984: 38–74) and Sarasin Viraphol (1977: 181–4) have shown that these monopolies were by no means unchanging nor necessarily a barrier to wider participation in trade. Indeed Evers (1987: 765) has argued that by the early nineteenth century the royal trade monopoly should be likened to the European trading companies. There is here a much wider point. Preoccupation with the continuity of forms or their similarity to European feudalism have blinded many writers to their very different and perhaps profoundly changing content.

Given that South East Asian production systems were by no means static on the eve of European penetration – and change appears to have accelerated with the rapid expansion of international trade after 1450 – what were they developing towards? A number of writers have concluded that in terms of organisation of the productive forces the Asian economies, including the major mainland South East Asian states, were in advance of Europe in the period 1300–1500. During this time the world economy may be regarded as comprising an interlocking system of trading spheres centring on major cities (Abu-Lughod, 1988: 2). Within these 'capitalistic institutions' had developed, for the extension of credit, the pooling of capital and risks and for sharing profits and losses. In addition production for export had begun to reorganise the way goods were produced and exchanged in the domestic economy. These developments were most profound in Asia and the Middle East (Abu-Lughod, 1988: 3). Blaut (1976) in describing the world *c.* 1500 refers to these cities as interconnecting nodes of 'proto-capitalism'. The implication being that until the expansion of the West the preconditions for the emergence of capitalism had appeared in many parts of the then world system. While these developments may well have been more advanced in Asia, after 1500 the balance shifted decisively in favour of

Europe. This is not to imply that the Asian economies stagnated between the penetration of the West during the sixteenth century and colonisation during the nineteenth, but rather that the productive forces were developed faster in Western Europe. Indeed, all the evidence points to considerable changes in South East Asia, particularly in those mainland states closely concerned with production for the international market.

Evers (1987) goes very much further, suggesting that by the early nineteenth century a form of 'peripheral capitalism' had emerged in Thailand. Additional weight is given to this view by the rapidity with which production for the world market expanded in Burma and Thailand after 1850 with apparently little change in the relations of production (see 4.2 and 4.3).

While a clear understanding of the pre-colonial relations of production that prevailed in pre-colonial South East Asia is necessary to any study of the incorporation of South East Asia into the world economy precise 'pigeon-holing' of the mode of production is of rather more academic interest. For many countries the weakness of the empirical information and the difficulties attendant on its interpretation makes it all too easy to fit the society into a preconceived formulation. This is particularly the case with the AMP because of the concept's general lack of precision.

Debates over whether these states were moving towards the preconditions for capitalism and, in the absence of the European transition, would have done so, seem of only limited utility. There is no inevitability about the transition from any one mode. The preconditions might emerge but the transition is a result of very specific conditions. The fact is that the European states made the transition first and became the core of a world economic system which effectively subjugated the South East Asian economies. It is the nature of this process that is most pertinent to the present study.

3

Western penetration: from trade to colonial annexation

3.1 South East Asian trade and Western penetration

The penetration of the Asian trading network by the Western powers was intimately connected with the rise of European capitalism. Thus the distinctive phases in the process reflect the development of the productive forces within those countries that were emerging as the core of the world-economy. The initial penetration by individual traders and, later, companies was the product of mercantile capitalism. This had neither the strength nor the imperative to move from control over South East Asia's external trade to control over territory and direct participation in production. Only in those areas lacking major political structures, for example the Philippines and parts of present-day Indonesia, was direct European control feasible before the nineteenth century. Not until the industrial revolution did the European nations become sufficiently powerful to dominate the major South East Asian states. This chapter sets out the general sequence of European penetration with the objective of illustrating the varied and changing nature of the processes that operated.

Prior to the penetration of the European powers South East Asia held a key position.[1] Trade networks connected East Africa with China and supplied the Middle East, Mediterranean and European world with a variety of high-value goods such as silk, spices, drugs and porcelain. While a wide range of commodities originated in South East Asia, by far the most significant were spices (map 3.1).[2] Pepper and ginger were grown over a fairly wide area, including India, Sri Lanka and many parts of South East Asia, most notably Sumatra; the more expensive cloves and nutmegs were restricted to the islands of Ternate, Tidore, Amboyna (Ambon) and the Bandas (Fisher, 1964: 128). The increased

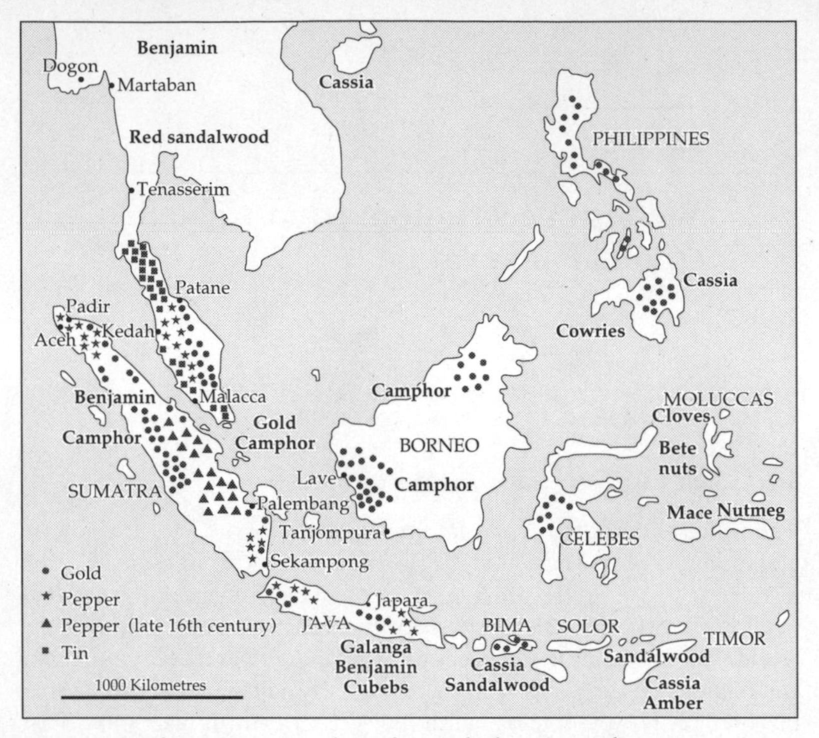

Map 3.1 South East Asian products during the late sixteenth century
Source: Braudel, 1979: 527

demand for spice in Europe[3] resulted in substantial increases in production during the fifteenth century. It has been estimated that by the end of that century perhaps 20 per cent of South East Asian production reached Europe (Wallerstein, 1974: 329). This traffic was firmly in the hands of Gujerati and Arab traders who exchanged, in particular, Indian textiles for spices. Production of cloves and nutmeg was firmly controlled by the small Malay Sultanates. However, from an early date spice production in the Celebes and Moluccas was a source of contention amongst the states of the islands (Harrison, 1963: 71).

In addition to its function as a supplier of commodities South East Asia, by virtue of its location, played a pivotal role in the Asia regional trade. The trade with China inevitably passed through the Straits of Malacca.[4] Control over the Straits formed the basis of the power of successive political units. From 1409 this power was held by Malacca[5] which during the fifteenth century provided a secure route through the Straits and grew into a major entrepôt. Tomé Pires[6] estimated that in

1510 2.4 million cruzados' worth of trade passed through Malacca and at any time there were over 2,000 ships at anchor in the harbour. Reid (1980: 239) puts this volume of trade in perspective by noting that in the *late* sixteenth century the value of goods entering Seville was about 4 million cruzados. Malacca was the key not only to the trade of South East Asia but to the Asian region as a whole.

By the beginning of the sixteenth century Malacca may well have had a population of 100,000 (Andaya and Andaya, 1982: 44). However, it was by no means the only major urban trading centre in the region. As Reid (1980: 325) has stressed 'the period from the late fourteenth to the seventeenth century . . . affords a picture of exceptionally rapid growth of trade and of networks of indigenous cities necessary to sustain it.' Estimates of city size are uneven and far from satisfactory, but all the evidence points to major centres being large by European standards, Ayuthia, Demak, Aceh, Makassar, Surabaya and Banten all being in the 50,000–100,000 range.[7]

The history and imperatives behind the initial penetration of Asia by European traders cannot be discussed in detail here.[8] That the process was intimately connected with the changing relations of production in what was to become the core of the world economy there can be little dispute. However, for the early voyagers and their backers, the immediate objective was the spice trade. Rising demand for spices in Europe, limited supply, the length and hazards of the trade route and the large number of intermediaries dictated that prices were extremely high. The operation of the Asian spice trade and the enormous profits that could be made from it were well known in Europe before a direct sea route existed. Indeed the Portuguese, in particular, appear to have set out to collect detailed information during the 1480s (Parry, 1963: 177).

Portugal and Spain

Following Vasco da Gama's voyage of 1497–9, the Portuguese established a commanding position in the Asian trading economy with remarkable rapidity. Goa was taken in 1510, Malacca in 1511 and Canton and the Moluccas reached in 1513. Essentially the Portuguese trading structure rested on a small number of often precariously held fortified trading posts (map 3.2) and the utilisation of their superiority at sea[9] to take trade out of the hands of the Arabs, Bengalis and Gujeratis (Boxer, 1969: 46). For a hundred years the Portuguese dominated the route from Asia to Europe and Lisbon became the 'spice capital of Europe'. Within Asia the Portuguese were never able to dominate the trading network,[10] but they established for themselves a

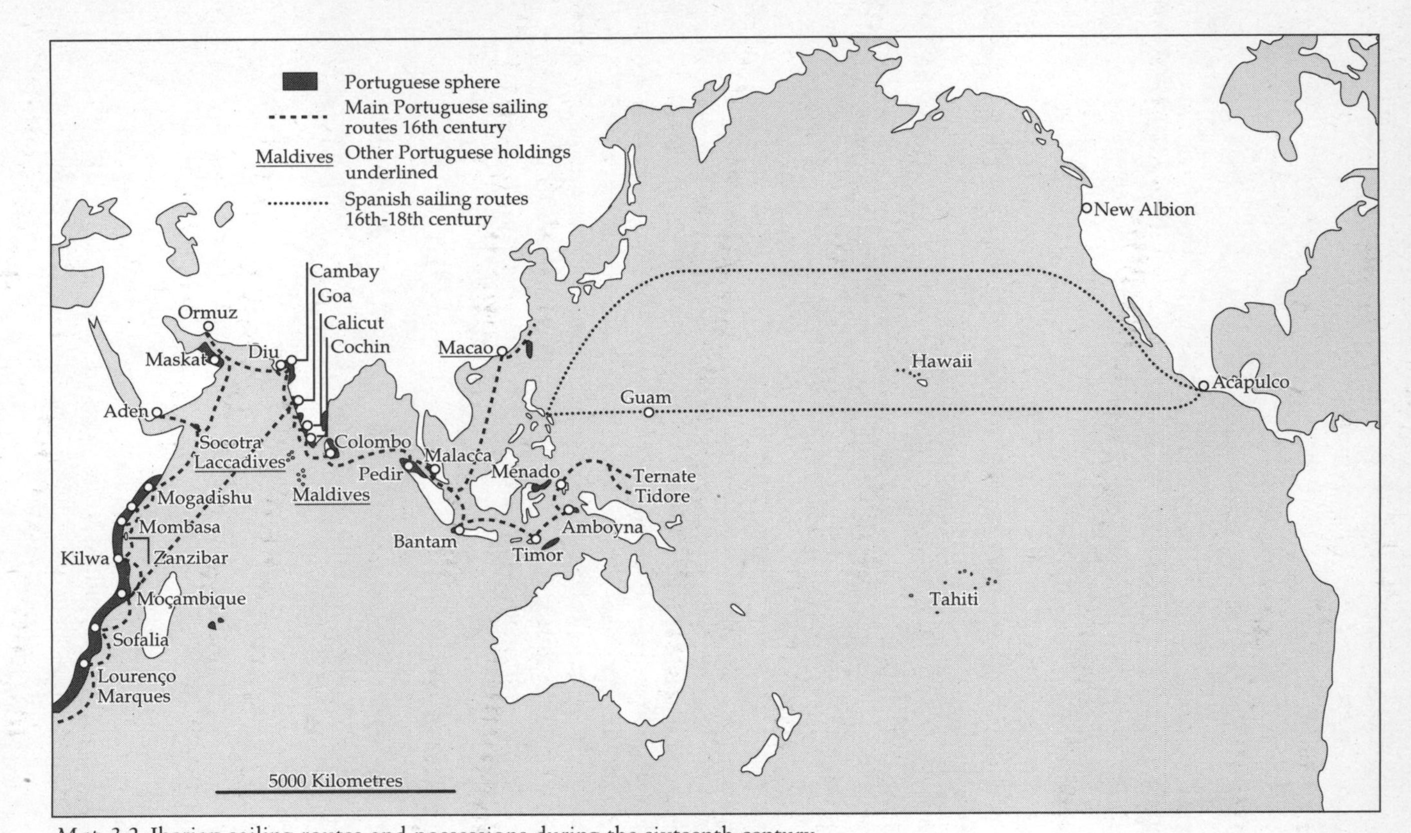

Map 3.2 Iberian sailing routes and possessions during the sixteenth century
Source: Fisher, 1964: 129, 132

wide-ranging trading role amongst the other long-established groups (Parry, 1963: 240).

While the main focus of Portuguese activity was the shipment of spice, there was little that Portugal could ship in return. Indeed, there was little that Europe as a whole could offer the Asian world. In consequence, spices were purchased rather than traded for (Boxer, 1969: 52). It is this aspect of the Portuguese activities that is basic to Wallerstein's (1974) assertion that in the sixteenth century Asia remained an 'external arena' to the European world economy whereas the Americas, which were involved in a two-way trade with Europe, were already a peripheral part of it. However, whatever the drain on bullion that it involved, the profits derived from the spice trade were enormous.[11] To provide trading capital the Portuguese merchants engaged in a wide range of ancillary trade throughout the Asian region. Indeed, it is the diversity and great geographical range of their activities that distinguishes the Portuguese from the other merchant communities. The records of these Portuguese activities tell us much about pre-colonial trade and manufacture in Asia as a whole:[12]

> The basis of this local commerce was the export trade in cotton textiles from the ports of Gujerat and Coromandel. These found a ready sale in Indonesia where they were exchanged for spices and in East Africa where they were bartered for gold and ivory ...... Equally profitable to the Portuguese was the acute demand in Japan for Chinese manufactured goods ...... There were many other local trades: the sandalwood trade for instance, between the China coast and the Lesser Sundia [Sunda] Islands; the slave trade ...; the import of horses from Mesopotamia and copper from Arabia, the export from India to China and Japan of hawks, peacocks and even the occasional tiger. (Parry, 1963: 240–1)

Thus the Portuguese inserted themselves into an established 'Asian World Economy' which they could in no way control. Indeed, until the industrial revolution the European powers in general were in no position to confront the major states of Asia (Boxer, 1969: 57; Wallerstein, 1974: 332). Portuguese territorial control was never to progress significantly beyond the scatter of ports shown in map 3.2.

From their arrival in Malacca in 1511 until the Spanish established themselves in the Philippines in 1564, the Portuguese had no European rivals in the region. The Spanish made limited attempts to establish themselves in the Moluccas but were rebuffed by the Portuguese and the local Sultans. Spanish trading became increasingly restricted to the Philippines. By the end of the sixteenth century Manila had become a major collection centre and a thriving trade in Chinese goods had been established. In effect the Manila trade was an extension of Spanish activities in the Americas. The silver-rich possessions of the New World provided the purchasing power for Chinese velvet, porcelain, bronzes, jade and, above all, silk.[13] This trade provided the main source of income for the Philippine colony. For 200 years the 'Manila galleon' route (map 3.2) linked China with Acapulco, and indirectly to Europe.

The Spanish occupation of the Philippines after 1565 was by far the earliest and in cultural terms probably the most complete in South East Asia. The islands were generally thinly populated (Fisher, 1964: 700) and, unlike the other areas of early European contact, they had little organisation above that of the village (Tate, 1971: 335–7).[14] Only in the southern islands of Mindanao, Palawan and the Sulus were there larger units, principally a product of the penetration of Islam during the preceding century. Thus a small force of Spaniards was able to establish itself in key areas between 1565 and 1573, following the path of direct conquest they had initiated in the Americas.

While the Spanish were able to establish a base in the lowlands of Central Luzon from which to subjugate the islands this task was scarcely completed at the end of their rule in 1898. The population of the mountainous areas of north western Luzon and the Sultanates of the southern islands provided a protracted resistance. Only in the mid-nineteenth century was the power of the latter finally broken.

The Philippines lacked spices, precious metals or any of the other valuable commodities that attracted the European powers to Asia. This, together with their remoteness from Spain and distant administration as part of the Vice Royalty of New Spain which centred on Mexico, goes far to explain the comparative neglect from which the islands suffered (Fisher, 1964: 701).

The development of the Philippines under the Spanish followed the pattern established in the Americas. Large lay and ecclesiastic estates were established. In order to defer the costs of 'Christianization and aid in the cost of maintaining Christianity, civil administration and protection' a system of tribute and labour service was imposed on the indigenous population (Juan de Solorzano Pereita, 1647, cited Cushner, 1971: 101). The labour levies were used to clear land and new crops, notably tobacco, maize, peanuts and potatoes were introduced. However, the islands attracted comparatively few *conquistadores* and land tended to accrue to the religious organisations (Fisher, 1964: 701). Thus the colonisation of the Philippines became dominated by the main religious orders, the majority of the towns growing out of mission centres.[15] With some truth have the Philippines been referred to as the 'empire of the friars'.

Throughout South East Asia Portuguese trading, military and missionary activities aroused great opposition. Malacca was subject to repeated attacks. Johor, Aheh and Java launched fifteen major assaults between 1513 and 1616. A variety of agreements were reached with the Sultans in the Moluccas over the supply of spice and a series of trading ports were established, most notably in Ternate. However, the Portuguese position remained precarious and in the islands as a whole they never became more than 'a new competitor in the old-existing pattern of political and commercial rivalry and power' (Tate, 1971: 45). As Harrison (1963: 71) has noted:

> In the Moluccas especially, where the chief rulers had accepted Islam and where the sources of the most highly valued spices were found, religious hatred and commercial greed sustained war and terrorism for many years. Warfare was already endemic in that area before the arrival of the Portuguese, for the Sultans of the two main islands of the group, Ternate and Tidore, were traditional enemies; but it was Portuguese policy to keep this warfare alive, promising assistance to one side or the other – or to both – for the price of commercial concessions.

By the late sixteenth century the comparatively weak and fragmented political structure which the Portuguese had first encountered was being replaced by a series of comparatively powerful states, most notably Aceh, Ternate and Mataram. The Portuguese activities almost certainly accelerated this process, firstly by acting as a focus for opposition forces and secondly, and probably more significantly, by increasing the opportunities for trade (Kirk, 1990; Tate, 1971: 47). In 1574 the Portuguese were expelled from Ternate and their power was generally weakened throughout the region (Parry, 1963: 93–97).[16] However, in the long run the most serious threat to the Portuguese was the weakness of Portugal *vis à vis* the other European powers and the increased interest of Dutch and English traders in the spice trade.

Holland

By the early 1600s England and Holland were emerging as the core of what Wallerstein (1974: 107) has called the 'European world economy'. The rapid rise in manufacturing, agricultural productivity and commercial activity within a general framework of changing relations of production in this core were closely tied to the growth of long-distance trade.[17] Indeed, as Rich (1967: xii) noted:

> A new world-economy was ... created [in the second half of the fifteenth century], an economy in which Lisbon and the Casa de Contratacion controlled the spice trade of the world and directed the fleet of spice-ships to their entrepôt

at Goa and then to the anchorages of the Tagus. Portuguese administration and financial techniques proved inadequate for such lucrative burdens [and] the Dutch proved their capacity as interlopers ... [T]he spice trade under Dutch control formed an invaluable adjunct to their trade to the Baltic and to northwestern Europe. The new and expanded trade in spices and eastern produce was geared into a trade system which spread throughout Europe and, indeed, across the Atlantic.

During the second half of the sixteenth century Amsterdam effectively created a framework for the smooth running of the world-economy (Wallerstein, 1974: 199). These developments cannot be discussed here, but the Dutch penetration of the South East Asian trade must be seen as part of this wider process. Dutch activity in South East Asia cannot be separated from the nationalist struggle against the Iberian powers[18] and the trading rivalry with England, the latter giving rise to the wars of 1652–4, 1665–7 and 1672–4.

During the sixteenth century the Dutch established themselves as the principal distributors of spice to northern Europe (Parry, 1963: 248). The closure of Lisbon to Dutch shipping, the 'revolt' of the Low Countries in 1566–79 and the growing Anglo-Spanish conflict seriously disrupted the European supply of spice. This stimulated northern European, particularly Dutch, traders to establish direct trade routes with South East Asia. At the same time the Iberian powers' trading empire in Asia could be attacked at its weakest and most vulnerable point.

The first Dutch trading expedition reached Bantam, a major collection centre, particularly for pepper, in 1596. Dutch activity expanded rapidly, the Moluccas were visited in 1599 and between 1595 and 1601 sixty-five ships in fifteen separate expeditions sailed for Asia (Harrison, 1963: 89). Like the Portuguese, the Dutch attempted to set up a system of local alliances which would provide trading concessions and cooperation against rival traders. In 1600 an agreement was reached with Amboyna (Ambon) which gave them a virtual monopoly of the nutmeg trade (Hall, 1981: 316). This provided Holland with its first important foothold in the islands. However, the Dutch encountered serious problems in enforcing agreements with local rulers and only firmly established their position in the Moluccas through a series of localised but often bloody conflicts.

In 1602 the various trading and military operations in South East Asia were unified under the Dutch East India Company (VOC). From 1610 when it became fully effective until its abolition in 1798 it controlled, with the force of the state, Dutch trading and administrative activities in the region. While it may be argued that during the seventeenth and

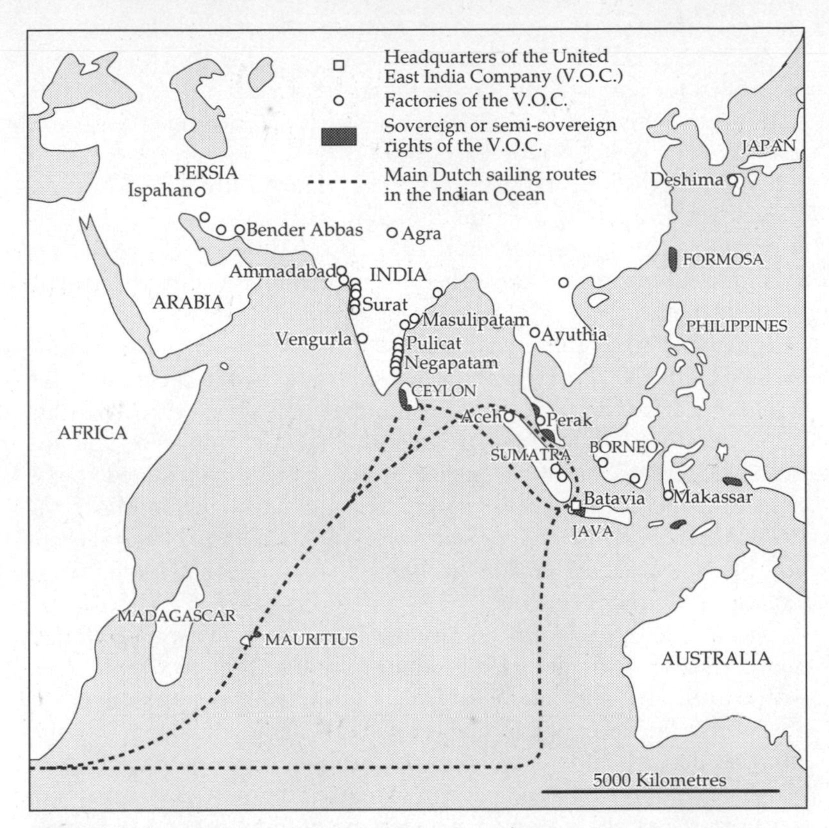

Map 3.3 VOC possessions and sailing routes during the seventeenth and eighteenth centuries
Source: Fisher, 1964: 132

eighteenth centuries the expansion of European trade and territorial control was a function of companies rather than the state, in South East Asia only the VOC played a major role.

Dutch trading organisations, resources and sea-power were much superior to those of the Portuguese (Tate, 1971: 48). What appears to have been a systematic war of attrition was conducted against the Portuguese, reducing their footholds and cutting off their supplies of spice. Malacca was starved of trade and the Dutch post at Batavia became increasingly central to the region's trade. By 1620 the trade passing through Malacca was perhaps only 50 per cent of that of the 1580s (Subrahmanyam, 1988: 75). This situation was to persist after the Dutch seizure of Malacca in 1641.

The focus of Dutch activity at Batavia reflected their interest in Java and their use of the more hazardous but more direct Sunda Straits route to the Moluccas.[19] With the capture of Malacca the Dutch effectively controlled the extra-regional trade routes (map 3.3).

While the Dutch were never able to entirely exclude other European traders from Island South East Asia, they seriously restricted their activities. English traders were effectively barred from the Moluccas after 1628. From 1685 until the founding of Penang in 1786 the only significant English presence was at Bencoolen. During this period the East India Company's activities became more than ever concentrated on India (Harrison, 1963: 124).

By the 1620s the Dutch had become the most powerful force in the islands. Unlike the Portuguese they were able to disrupt the trade to the extent that from this period 'the great indigenous states withered away' (Tate, 1971: 48). The gradual advance of Dutch influence and the successive crumbling of the major political units contributed to the general fragmentation of the islands and the Malay peninsula (Tate, 1971: 94). This not only kept states weak and facilitated later annexation but had a lasting impact on the post-colonial countries that were formed out of the fragments.

Until the late eighteenth century the Dutch in many ways followed the pattern of the earlier island trading empires. They were concerned with controlling trade and establishing monopolistic powers over the producers. However, their policy of establishing trading posts and alliances drew them inevitably into controlling territory (map 3.4). As Harrison (1963: 152) has noted: 'On Java itself the Dutch Company was transformed, as was its British counterpart in India, from merchant to landlord.' This trend was accelerated by conflicts with the major states[20] of Bantam, Macassar, Mataram and Aceh and the fears of local alliances with rival European trading powers.

The mounting costs of administration and major military campaigns combined with increasing penetration of the Dutch monopolies by other European traders to steadily reduce the profitability of the VOC during the eighteenth century. This tendency was closely bound up with the eclipse of Holland as a major European power. By the time of its abolition in 1799 the VOC was effectively bankrupt.

Britain

During the eighteenth century British EIC (East India Company) and private merchant trade with China increased steadily. Many of the private merchants went no further than South East Asia if they could exchange their cloth and opium for spices (Harrison, 1963: 143). These

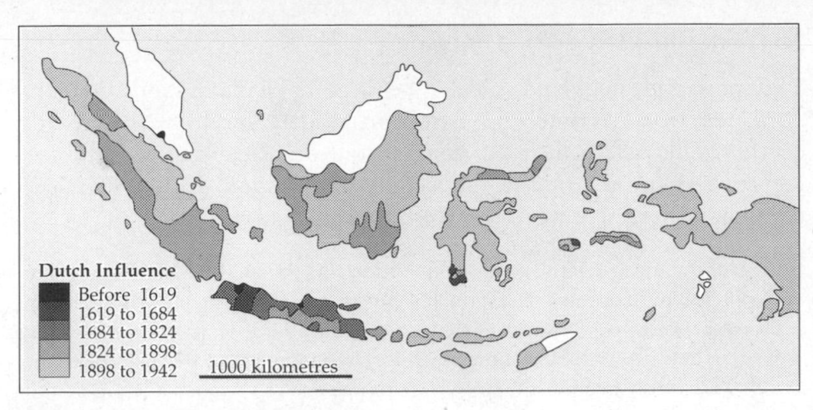

Map 3.4 Expansion of Dutch territorial control
Source: Fisher, 1964: 257

voyages increased as Dutch power waned, a process greatly accelerated by the war of 1780–4.[21] This greatly weakened Dutch commercial and naval power, and in the peace treaty the right of English merchants to trade throughout South East Asia was formally recognised.

England emerged from the wars of 1778–84 as the premier naval, trading and manufacturing power. A period of rapid growth of overseas trade ensued while the

> balance of Britain's overseas interests and power was now shifting from North America to Asia, at a time when her expanding industries were demanding an increasing supply of tropical raw material ... Asia offered a vast potential market for a country possessing, as Britain did, world leadership in the production of cheap manufacturing goods. With her Manchester cottons, as well as with her Indian opium, Britain could be in a position to build up prominent control of the markets of India, South East Asia and the Far East. (Harrison, 1963: 155–6)

Sporadic attempts were made by the EIC during the eighteenth century to establish a base other than Bencoolen to act as a way-station for the China trade.[22] The rapid expansion of trade after 1784 gave added impetus to their endeavours, which culminated in Francis Light's occupation of Penang in 1786. This base was to become of major importance as a base for British military operations against the Dutch territories during the wars of 1793–1815.[23] The Dutch territories occupied by Britain during these conflicts were all returned to Holland in 1816 but the trading and naval power of Holland was further reduced and the basis laid for major British penetration later in the nineteenth century.

Mercantilism and the mainland states

The powerful mainland states of Burma, Thailand and Vietnam remained peripheral to the European traders until the nineteenth century. This reflects their location, strength and lack of commodities as lucrative as those found in the islands. In addition, Hall (1981: 339) could have been writing of the Western traders in general:

> Wherever there was relatively fair competition, the Asian, Arab, Persian, Indian or Chinese could always maintain his position. Only where the Dutchman could resort to force, as in the Spice Islands, could he gain advantage over the Asian traders ... With the more powerful monarchies of the mainland the Dutch were rarely in a position to dictate terms, and the Asian traders were too well established to be ousted.

However, although the intensity of activity varied, there was a continuous European presence from the sixteenth century and some premature attempts to establish territorial control.[24]

The Portuguese were trading with Ayuthia from 1512, introducing fire-arms, cannons and European-style fortifications (Tate, 1971: 501). The Dutch established a factory there in 1602 and by 1617 had gained a virtual monopoly of the hide trade. In 1644, following a brief armed confrontation, they imposed an unequal treaty which aimed at establishing a Dutch monopoly over the country's international trade (Sarasin Viraphol, 1977: 9–15). In what appears to have been an attempt to balance Dutch influence, the EIC and later the French were made welcome. In the event the French were to prove a greater threat than the Dutch. In 1687–8 they made a determined effort to oust the Dutch and effectively annex the kingdom.[25] The attempt failed, but the episode left the Thais with an extremely cautious attitude to the Western powers which persisted until the nineteenth century (Tate, 1971: 503).

The other mainland states, also, from the sixteenth century had their share of missionaries, traders and adventurers, some of whom tried to establish territorial bases, particularly along the Tenasserim coast. However, none of the European powers succeeded in establishing a permanent presence in Burma despite its strategic importance for the Anglo-French rivalry in the Indian Ocean and its resources of teak, gold and precious stones. Missionary activity, particularly by the French, was a significant element in Western penetration of the mainland. Their work was frequently curtailed and the cause of friction; this was, at least in part, due to their interference in internal politics. In terms of long-term development, missionary activity was most important in Vietnam. By the 1750s the number of Vietnamese Christians was estimated at

300,000 (Tate, 1971: 442). The intensification of missionary activity during the nineteenth century and the increased conflict with the Vietnamese state that resulted was to give a pretext for French intervention.

By 1800, while there was a European presence throughout the region, only in parts of present-day Indonesia and the Philippines had territorial control been established. As a direct result of the Western intervention the states of the islands and Malay Peninsula were declining and fragmenting. While the mainland states were strengthened by the growth of trade the significance of the presence of Western traders, missionaries, envoys, advisers and adventurers is much more difficult to assess. It is probable that these contacts were important in forming the various attitudes towards European powers that prevailed during the nineteenth century. Perhaps more significantly these states were also receptive to European ideas and technology. Most comment has been reserved for the adoption of Western weapons[26] and new crops; little attention has focused on other introductions. Yet in Thailand, Western shipbuilding techniques and designs were being incorporated into the indigenous industry from the late eighteenth century.[27] It is unclear how widespread these developments were or whether their adoption reflected changes in the relations of production. However, the early nineteenth century was a period of comparatively rapid change in Burma, Thailand and Vietnam. The significance of the European contact in this has received comparatively little attention.[28]

3.2 From trade to colonisation

In the course of the nineteenth century South East Asia passed firmly into European control. By 1909 only Thailand, much reduced in size, retained national independence (map 3.5).

This change was both a reflection and part of the development of international capitalism. The European states that annexed South East Asia in the nineteenth century were very different from those whose traders had earlier competed for the region's commerce. It was not merely that the European powers in general, and Britain in particular, had become sufficiently strong that annexation was a practical option; they had also undergone a qualitative change which made such a development necessary.

This is not the place for a debate over the imperatives that lay behind nineteenth-century colonisation. It is accepted here that the process was an integral and necessary part of the development of capitalism. As such we are looking at a process operating at the global scale. However, within this evolving global system the role of Britain was crucial.

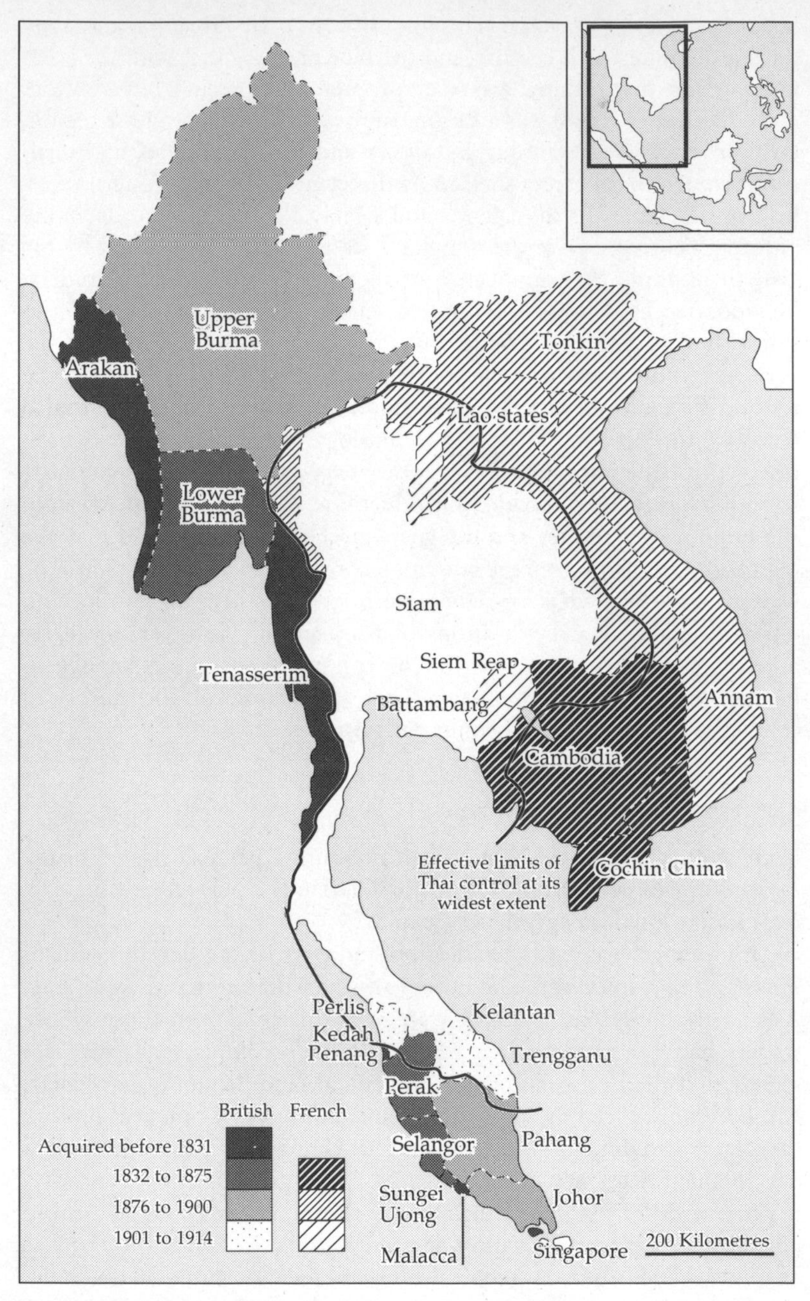

Map 3.5 Expansion of European control
Source: Fisher, 1964: 150; Tate, 1971: 507

Britain

During the early part of the nineteenth century Britain had emerged as the leading national power with a virtual monopoly in industrial production. This strength fostered the doctrine of 'free trade' and the extension of British interests 'informally through a vast commercial network rather than by direct colonial rule' (Murray, 1980: 11). This was the pattern followed in South East Asia during the first three-quarters of the nineteenth century. Territorial annexation was the exception and may be regarded as a 'strategy of last resort when informal methods had failed to provide security for British interests' (Gallagher and Robinson, 1953: 9–11).

During the last quarter of the nineteenth century the situation was radically transformed: territorial annexation became the norm. The challenge to Britain's leading role in the world-economy that resulted from the diffusion of the industrial revolution to the rest of Europe and North America heralded a period of intense competition.

In the earlier phase of annexation the protection of established commercial activities was paramount. From the 1870s annexation increasingly took place ahead of trade and investment. In essence the world was sub-divided into a series of 'reserves' for the future activity of individual metropolitan countries' capital. The pace of activity and the need to outmanoeuvre rivals resulted in many annexations taking place with little accurate information on their value. As a result 'the imperialist powers were frequently content to allow local economic activities to stagnate rather than allow a rival metropolitan state administration to assume either formal or informal control' (Murray, 1980: 13).

At the level of immediate causes a variety of factors governed individual annexations. These included retaliation, strategic considerations, protection of traders and missionaries, ambition and even the spreading of Western civilisation and religion, but such factors do not explain late-nineteenth-century colonisation.

In South East Asia the process of colonisation was accelerated by the rapid growth of Western trade and investment following the opening of the Suez Canal in 1869 and the establishment of telegraph communications from 1871. The region became effectively closer to the Western industrial centres at a time when the search for markets and the demand for the region's raw materials were increasing rapidly.

European activity in South East Asia was further stimulated by renewed interest in the region's strategic location for trade with China. This was particularly important in the growth of British interest in South East Asia. Sar Desai (1977: 23) goes rather further:

Until the middle of the nineteenth century British interest in South East Asia was linked to that region's strategic location between India and China. Up to that point of time Britain had centred her commercial activities in the East upon India and conditions permitting, on China. The Portuguese, Dutch, French, and English had all tried to establish trading relations with China in the previous two centuries but with very limited success. After the advent of the Industrial Revolution with prospects of burgeoning surpluses of manufactured goods, especially textiles, British interest in opening Chinese markets intensified, because China's enormous population seemed to offer unlimited possibilities of commercial expansion. Correspondingly, British interest and involvement in the exercise of control over trade routes to China, both maritime and overland, manifested themselves all through the nineteenth century. South East Asia's strategic location in the context of the potentially lucrative China trade revived British interest in a region which had been practically abandoned by the East India Company to the Dutch after the 'massacre' on Amboyna Island in 1623. If Britain's emphasis in the first half of the nineteenth century was upon exploitation of the sea route and, therefore, in securing strategic positions along the Straits of Malacca and the South China Sea, the second half witnessed a growing preoccupation with exploring and monopolizing a land route across Burma as a backdoor to South-West China. The expansion of British Empire in South East Asia must be examined against such a backdrop of the development of Anglo-Chinese commercial activity.

The 'backdoor' to China was perhaps the most important factor, particularly in the intensification of Anglo-French rivalry over the control of mainland South East Asia.

The rapid increase of EIC and private trade with China was, because of the lack of British goods acceptable to the Chinese, matched by a steadily worsening balance of trade.[29] Trade with China was increasingly balanced by the export of Indian opium and cotton and, to a lesser extent, by British participation in South East Asian trade. The latter took two forms. Firstly trade within the region which supplied cash to purchase Chinese goods; and secondly the acquisition of South East Asian commodities which were acceptable in China. Neither of these activities could be easily expanded without impinging further on the Dutch interests or the regional trade that remained largely in Chinese hands (Sar Desai 1977: 27–8).

A wide range of South East Asian products – rattans, pepper, tin, birds' nests, betel-nut, coral, amber, sandalwood, spices and rice – had long been traded with China. This growing trade was largely in Chinese hands and for much of it Bangkok acted as the entrepôt (Sar Desai, 1977: 26). The establishment of Penang in 1784 had, however, given Britain an important foothold in the region's trade, despite the locational disadvantages of the port.[30]

The founding of Singapore in 1819, and its establishment as a free port, marked a decisive step in Britain's penetration of South East Asia. Advantages of location, harbour, lack of tariffs and security rapidly attracted trade from other countries.[31] By 1824 Singapore was the most prosperous entrepôt in the islands and by 1834 had supplanted Bangkok as the collection centre for South East Asian produce destined for the China market (Wong Lin Ken, 1978). From the 1870s Singapore became central to the region's trade and communication – a reflection of Britain's preeminent position in world finance, trade and manufacturing.

Singapore's strategic and commercial importance was greatly enhanced by the increased interest in the China trade following the 'Opium War' of 1839–42 and the establishing of Hongkong in 1841. However, the opening of the five Treaty Ports in 1842 failed to generate the level of trade growth expected (Sargent, 1907: 133). This focused increased attention on the expansion of South East Asian trade as an outlet for British goods and, through the regional trade with China, as a means of generating purchasing power in China (Sar Desai, 1977: 55).

The strength of British capitalism served to spread trading activities widely through the islands, Malay Peninsula and, to a lesser extent, the mainland states of Burma, Cambodia and Vietnam. These activities resulted in increased friction with the Dutch and French. The Anglo-Dutch treaty of 1824 recognised the Malay Peninsula as a sphere of British influence and the islands south of the Straits of Malacca as a sphere of Dutch activity. While this treaty paved the way for British control of the Malay Peninsula and the consolidation of the Dutch position in the islands it did not bring an end to the disputes over the island trade. The British, in particular, continued to have a stranglehold on the shipping within the NEI, while in areas outside of formal Dutch control, for example in Atjeh and northeast Sumatra, British traders remained dominant until the late nineteenth century (Tate, 1971: 226–31). However, the progress and consolidation of Dutch control (map 3.4) gradually reduced the areas of British activity. In the 1890s the establishment of the KPM (Koninglijke Paketvaart Mij) shipping line substantially reduced British activity in the islands and diverted much NEI trade away from Singapore (Tate, 1979: 156).

The penetration of British trade into the Malay Peninsula was accompanied and at times facilitated by a whole range of informal agreements and relationships. Commercial interests repeatedly lobbied for the extension of British control in order to protect and facilitate the expansion of their activities. This was particularly so during the 1860s when the Singapore merchants were faced with declining trade as a result of

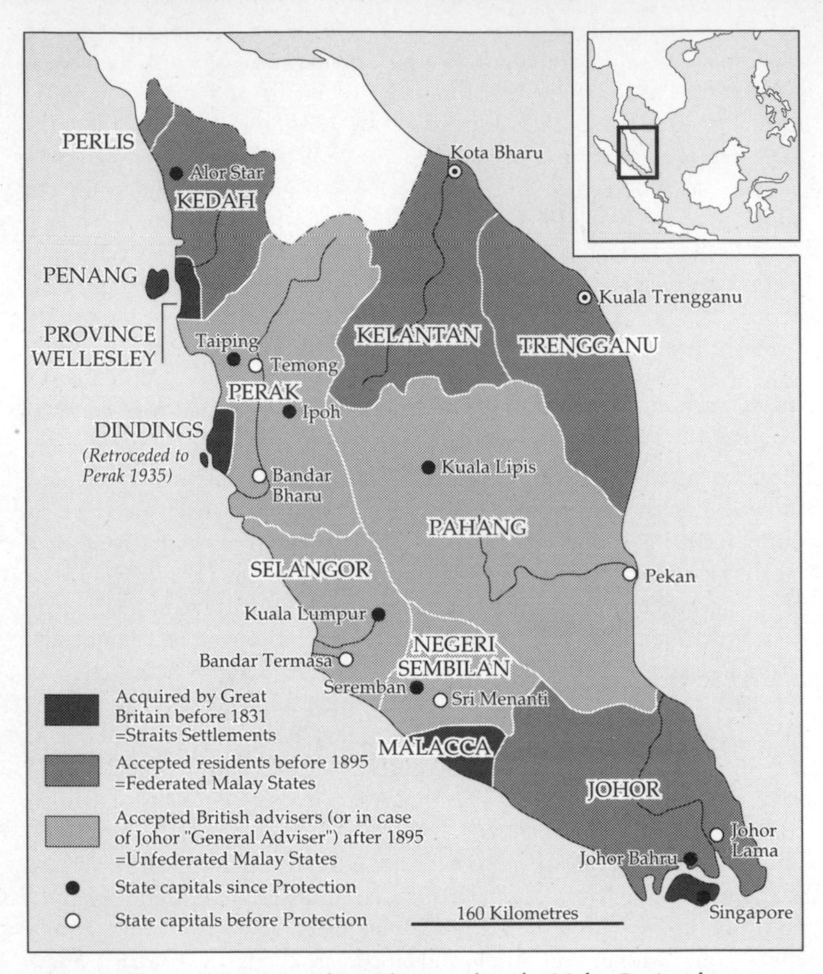

Map 3.6 The establishment of British control in the Malay Peninsula
Source: Fisher: 1964

increased competition from the other Western powers.[32] The way was opened to a 'forward policy' in the Malay States by the establishment of the Straits Settlements in 1867.

Between 1874 and 1914 the Malay Peninsula passed into a complex pattern of direct and indirect rule (map 3.6; Cowan, 1961). The rapid expansion of tin mining following the discovery of the rich Larut field in 1848 was of major importance in attracting Chinese and British capital to the Peninsula. The rapid influx of Chinese miners was accompanied by increasingly prevalent disputes with the Malays and between different groups of Chinese (Fisher, 1964: 596). British commercial interests became increasingly vocal in their demands for protection of their interests and the establishment of 'law and order'.[33] However, while this provided the ostensible purpose of British intervention, it has to be seen in the broader context of the wider European struggle for control in South East Asia following the opening of the Suez Canal in 1869 (Parkinson, 1960: 16). In 1874 protectorates were established over the main tin-producing states of Perak, Selangor and Sungai Ujong (the key area of Negeri Sembilan, the remainder of which became a protectorate in 1889). All these states lay to the east of the Main Range with the majority of their tin-mining areas within fifty miles of the coast. In contrast Pahang, which became a protectorate in 1880, lay to the east and contained no important tin-mining areas, though it was at the time believed, incorrectly, to be extremely rich in gold (Fisher, 1964: 597). In 1895–6 the four protected states were united as the Federated Malay States with the capital at Kuala Lumpur. From this date the administration became increasingly centralised and the powers of the states notional. The establishment of British rule did not take place without opposition from the Malays, particularly in Perak (Hall, 1981: 607).

Johor, because of its lack of mineral resources, had received less attention and was to come under much looser control after 1885. From 1909 it was grouped with the states of Kedah, Perlis, Kelantan and Trengganu, ceded in that year from Thailand. They refused to join the Federation and were left with a greater degree of authority, particularly over finance (Hall, 1981: 607).

In Borneo a complex and in many ways bizarre series of developments between 1846 and 1888 established British control. From 1846 Sarawak was ruled over by the Brooke family as an almost feudal holding. During the 1870s a variety of attempts by American, Austrian, Dutch and British agents to establish a presence in North Borneo culminated in the establishment of the British North Borneo Company. In 1888 fears that the weak sultanate of Brunei might seek an alliance with another Western power prompted the establishment of British 'protection' of Sarawak and British North Borneo.

By 1914 the British had established a remarkably varied and complex pattern of control in the Malay Peninsula and Borneo. Economic activity was concentrated in the Straits Settlements and the Federated State. In contrast Sarawak, Brunei, British North Borneo and the eastern part of the Peninsula appear to have been largely regarded as 'reserve areas' for future expansion (Fisher, 1964: 672). The pattern established by the British has left an indelible imprint on the successor states.

The expansion of British power into the Malay Peninsula and Borneo only encountered resistance from weak and fragmented states. In contrast, growing commericial and strategic interest in the Arakan coast[34] was to bring Britain into conflict with one of the most powerful states in the region (Tin Hla Thaw, 1977: 187). From the 1780s Burma had undergone a period of remarkable growth and territorial expansion.[35]

The spread of Burmese power into Arakan and Assam brought direct contact with areas of India under British control. The Anglo-Burmese war of 1824–6 was a result of disputes on these frontiers. However, the humiliating peace treaty of 1826 and the annexation of Arakan (map 3.5) opened the door for a variety of commercial interests. Traders were attracted by the potential markets, rich resources of teak and the possibilities offered by the overland trade with China. Despite the ineffectiveness of the commercial treaty of 1826, British commercial activity expanded rapidly most notably in Rangoon where a small but vigorous and increasingly vocal commercial community was established (Sar Desai, 1977: 118–34).

While the 1826 treaty did not impinge on Burmese sovereignty it did set in motion a pattern of commercial development which was ultimately to lead to the complete annexation of the kingdom. The acquisition of Lower Burma in 1852 and Upper Burma in 1886 were principally aimed at protecting and furthering British commercial activities (Singhal, 1960: 83). The extent to which annexation of Upper Burma was hastened by the signing of the Franco-Burmese Treaties of 1873 and 1885 and the fear of French intervention is a matter of considerable conjecture (Chew, 1979; Khoo Kay Kim, 1977).

The annexation of 1852 left Upper Burma as a land-locked state. All sea-borne trade had to pass through British-controlled territory. This proved a powerful weapon in subsequent negotiations. Following the commercial treaties of 1862 and 1867, British exploration (particularly of the routes to China), trade and exploitation of the teak resources of Upper Burma progressed with little hindrance (Tin Hla Thaw, 1977: 198–9). The spread of British influence following the 1867 treaty gave a certain inevitability to the annexation of Upper Burma in the climate of imperial rivalry that prevailed in the last quarter of the nineteenth century:

> Thus, annexation of Upper Burma in 1886 did not require the excuse of the unfriendliness of [King] Theeban, the unsatisfactory handling of teak contracts, or even the potential rivalry of French traders. These were important factors influencing timing, but they were not the cause of the annexation. (Woodman, 1962: 186)

The resistance to the British annexation and subsequent abolition of the monarchy was considerable. By mid-1887 there were over 40,000 British troops in the country. Resistance continued in Lower Burma until 1889 and in the hill states until 1895 (Tate, 1971: 420). Indeed Aung-Thwin (1985: 247) argues that British 'pacification' of Burma was not completed by 1948.

The annexation of Upper Burma did not bring into full British control the hill states of the northeast and northwest. These had enjoyed varying degrees of independence depending on the strength of the Burmese state. Following the annexation of Lower Burma British interest was attracted to the Karen states by their teak, minerals and strategic importance. During the 1870s the Burmese were forced to concede their interests in the area (Sar Desai, 1977: 186–90). These, like the other 'hill states', despite the importance of their resources to the imperial economy, were never fully incorporated into the formal colonial structure. In the long term this was to pose a major problem for the post-colonial Burmese state. Similarly the boundary between Burma and China was never fully demarcated, giving rise to boundary disputes that were not settled until 1960 (Fisher, 1971: 289–95).

France[36]

While the British were consolidating their position in Burma and the Malay Peninsula, French interests were being advanced in the eastern part of the mainland. Between 1862 and 1908 the whole of present-day Kampuchea, Laos and Vietnam passed into direct French rule. France had long-standing interests in missionary and trading activity in these areas. However, by the 1850s they were increasingly preoccupied with the establishment of colonies to replace the territory lost in India and to secure a route to China via the Mekong or Red rivers.

Annam, Cochinchina and Tonkin had attracted the attentions of European traders since the sixteenth century, particularly because of the proximity of these territories to China. All attempts to establish trading posts and commercial relations had been comparatively short-lived. During the eighteenth century sporadic attempts to establish regular trade were made by the French and British. However, given the difficulties of establishing relations with these inward-looking and far from stable states, and the preoccupation of Britain and France with India and North America, the main European activity centred on missionaries. French missionaries became closely involved with the limited trading activities that developed and, more importantly, acted as advisers to the local rulers.

Between 1777 and 1802 French influence increased rapidly. During this period the separate territories of Annam, Cochinchina and Tonkin were united by Nguyen Anh, forming a political unit similar to present-day Vietnam. French advisers and military volunteers played a significant part in this protracted and bloody process. However, the global conflicts of 1793–1815 and the British naval supremacy prevented France from capitalising on the position. Not until 1817 did French merchants re-establish trade with Vietnam. However, French influence was short-lived. Following the death of Nguyen Anh in 1820 the increasingly powerful and centralised Vietnamese state became rapidly anti-western. European trading and missionary activities were progressively restricted and Christian converts persecuted. French attempts to establish formal commercial contacts in the 1820s and 1830s were rejected.

During the 1820s and 1830s Vietnamese trade with Singapore increased rapidly. Goods were largely carried in royal Vietnamese and Cochinchinese vessels (Tarling, 1975: 94). While Britain generally maintained cordial relations with Vietnam only limited and unsuccessful attempts were made to obtain commercial treaties. Britain's increasing preoccupation with Burma, Thailand and the Malay Peninsula left little scope for expansion into Vietnam. Despite some misgivings amongst British traders and representatives that opportunities were being lost, and a French presence in the eastern part of the mainland might threaten the positions in China and Thailand, France was left with an increasingly free hand (Tarling, 1975: 92–133).

Events in Burma and China served to confirm the Vietnamese rulers of the wisdom of their anti-Western stance. During the 1840s the imprisonment of missionaries resulted in a number of military clashes with the French, most notably in 1843 and 1845. The opposition to Western contact became notably more marked after Tu-Duc became Emperor in 1848. During the 1850s Christian communities were dispossessed, large numbers being executed, branded or imprisoned. The execution of a number of missionaries during 1851–7 gave France an excuse to intervene directly in Vietnamese affairs.

French territorial ambitions in Vietnam have to be seen in the context of the opening of the trade with China. Vietnam's proximity to China, the availability of Chinese goods and the possibility of establishing an overland route to central China through Yunnan were major attractions for France. However, France's ability to intervene was limited by the weakness of French capitalism, the instability of the successive regimes, the need to cooperate with the other Western powers and the danger of antagonising China.

In 1859, following the seizure of Canton by an Anglo-French force, Cochinchina was invaded by a French force supported by Britain and Spain.[37] Tourane and Saigon were occupied but the campaign was interrupted by the renewal of hostilities with China. In 1861 a major French invasion force overcame considerable resistance to establish control over lower Cochinchina. These provinces were ceded to France under treaty in 1862. However, France faced continued resistance, a widespread 'rebellion' breaking out, in the suppression of which the remainder of Cochinchina was occupied.

In contrast to this bloody episode the imposition of protectorate status on the weak kingdom of Cambodia was a peaceful affair, primarily involving negotiations with Bangkok which claimed suzerainty over the kingdom. This was settled by the treaty of 1867 which confirmed Bangkok's control over the Cambodian provinces of Battambang and Angkor.[38] However, during 1866–7 French troops were heavily involved in the suppression of a major rebellion, which centred on an attempt to depose King Norodom. With the acquisition of Cambodia, France was well on the way to acquiring a major territorial base in South East Asia.

The establishment of French control over the Mekong delta opened the way to the exploration of the possible 'backdoor' routes to China. Expeditions in 1861 and 1868 led to the conclusion that the only practical route was that of the Red River. This focused French attention on the Tonkin area.

During the 1860s and 1870s Tonkin was in a state of chronic revolt and scarcely under the control of the Vietnamese government. France was constrained from any immediate direct action by the Franco-Prussian war and its aftermath. Additionally, as a hedge against further French territorial ambitions Vietnam sought closer ties with China, renewing 'tribute' payments and acknowledging 'vassal' status. Thus direct action in Tonkin was likely to precipitate a conflict with China.

In 1874, following an unauthorised attempt to seize control of Hanoi and Tourane, a wide-ranging treaty was signed.[39] On paper the gains for France were considerable: French sovereignty over Cochinchina was confirmed; the ports of Qui-nonh, Tourane and Hanoi were opened to French traders; navigation of the Red River was guaranteed; French officials were to be appointed to the Vietnamese customs service; and France gained 'most favoured nation status'. In practice France obtained little advantage from this treaty – the lack of effective Vietnamese control over the Red River rendered it unsafe for navigation. France, however, took no further immediate action.

The occupation of Tonkin in 1881 was made possible by the recovery of France from the effects of the Franco-Prussian war. However, the timing was governed by the fear that Britain would open a route to Yunnan through Burma (Hall, 1981: 766). The imposition of French rule was a bloody and protracted affair. Apart from the resistance of the Vietnamese there was open conflict with China which lasted until 1885. Not until 1895 was Tonkin considered to be 'pacified'.

Anglo-French rivalry over Thailand

Rival British and French penetration of mainland South East Asia gradually impinged on areas that were either under Thai control or owed allegiance to Bangkok. The situation was exacerbated by Thai recovery from the disaster of the Burmese invasion of 1760–6 and establishment or re-establishment of control over the northern Malay states and parts of Laos and Cambodia during the 1820s. Indeed during the first half of the nineteenth century the 'Siamese empire . . . was more powerful and extensive than at any previous time. It dwarfed all its main South East Asian neighbours in its sheer size and set an example for them by its ability to act constructively and forcefully in a dangerous world' (Wyatt, 1982: 180).

While Thailand was a major centre of South East Asian trade European participation in this was slight. It is often asserted that the Thai rulers' caution in dealing with the West stemmed from France's abortive attempt at annexation during the 1680s. Whatever the truth of this, during the early nineteenth century events in neighbouring states provided ample opportunity to study the methods and consequences of Western penetration.

Between 1818 and 1822 Britain made several unsuccessful attempts to establish formal commercial relations with Thailand. The 1826 Burney Treaty, although it was to govern Anglo-Thai relations until 1855 left trade (from a British point of view) seriously restricted. It did, however, go some way towards clarifying British and Thai spheres of influence in the Malay Peninsula. Thai control of Kedah and British control of Penang were confirmed. The situation in Perak and Selangor, however, remained ambiguous.

The Burney Treaty itself does not appear to have led to any substantial increase in direct Anglo-Thai trade. However, Thailand's international trade and general contact with Western powers increased markedly. In particular there was a rapid expansion of trade with Singapore (Wyatt, 1982: 169–70). From the late 1830s Thailand began to restrict trade. Duties on shipping were increased in 1839 and the royal monopolies were used to reduce European trade (Sar Desai, 1977:

84). It is likely that these represented a concerted attempt to divert trade into Thai-Chinese hands. During the 1840s British trading interests began to demand a new and more favourable treaty (Tarling, 1975: 134–173).[40]

By 1850 a situation was emerging where Thailand would either have to make major concessions to the West or engage in confrontation. The accession of King Mongkut (Rama IV) marked a decisive change in Thai relations with the Western nations.[41] In 1851 trade restrictions were eased and in 1855 the decisive Bowring Treaty was signed.[42] This established extra-territoriality, free trade and fixed import and export duties. Effectively Thailand surrendered control over customs duties and thereby over a large proportion of state revenue. Thailand became in effect an informal British colony.[43] In many respects the kingdom became an extension of the British colonial economies of the Malay Peninsula and Burma.

The concessions that Thailand was forced to make to the West did not end with the 1855 treaty. As may be seen from map 3.5 substantial areas of the kingdom were ceded to France and Britain. The former obtained the Lao states in 1893 and Siem Reap in 1907, while Britain received the northern Malay states in 1909. Indeed the kingdom was to be increasingly an object of Anglo-French rivalry (Goldman, 1972). During the 1860s King Mongkut aptly summarised the situation:

> Since we are now being constantly abused by the French because we will not allow ourselves to be placed under their domination like the Cambodians, it is for us to decide what to do; whether to swim up-river to make friends with the crocodile [the French] or to swim out to sea and hang on to the whale [the British] . . . It is sufficient for us to keep ourselves within our house and home; it may be necessary for us to forego some of our former power and influence. (cited Moffat, 1961: 124)

Anglo-French tension over Thailand was finally resolved in the treaties of 1896 and 1904 which effectively established the kingdom as a 'buffer zone'. While these treaties can be read as guaranteeing the independence of Thailand, in reality they reserved the kingdom as a field of operation for British capital (Harrison, 1963: 211; Goldman, 1972).

The rulers of Thailand were remarkably apt at exploiting the conditions presented by the expansion of British and French imperialism. However, as Tate (1979: 497) has noted:

> The rulers of the Chakri dynasty succeeded in their primary object, that of preserving the political sovereignty of Thailand, but this did not prevent – in fact was largely achieved by – the country becoming to all intents and purposes an economic 'satellite' of Great Britain in particular and an extension of the colonial imperiums of the West in general.

Outside of the fiscal, financial and trading spheres Thailand retained a degree of political independence. However, to a considerable extent, the independence was preserved by refraining from exercising it. Jacoby (1961: 55) indeed suggests that Thailand's position of partial political independence and economic subservience gave the worst of both worlds. Thailand did not benefit from the limited welfare development that colonial rule might have brought, nor was the kingdom given the benefit of the British Imperial protection from the worst of the fluctuations in the world-economy. This remains an area of considerable debate. However, the significant issue is that Thailand's unique (in South East Asia) process of incorporation has had a profound impact on subsequent development.

The USA and the Philippines

While the 'outburst' of 'new colonisation' by the European powers which characterised the last quarter of the nineteenth century was stimulated in part by the American industrial challenge, it also encouraged the USA to follow suit (Agnew, 1987: 59).

American merchants had been trading with China from 1784 and, during the nineteenth century, began to trade widely in South East Asia. These activities resulted in increasing friction with the European powers and repeated attempts were made, notably by the East India Company, to restrict American traders. In response to these difficulties the USA established a network of consulates to protect American commercial interests. However, as Britain, France and Holland consolidated their territorial control over South East Asia, American traders, like those of the other European powers, found their activities increasingly restricted.

To a degree the growing strength of American capitalism enabled continued penetration of the South East Asian markets, largely through the use of agencies and the establishment of subsidiary companies. As Kiernan (1974: 124) has noted, the USA was able to expand by 'employing the earlier empire-builders as managers'. However, in order to establish an effective position in the Asian region, an element of territorial control was necessary. Between 1856 and 1899 the USA 'island hopped' across the Pacific, establishing a chain of bases from Hawaii to the Philippines. In this respect, by the 1890s the needs of the large American firms and the policies of the government of the USA were conjoined (Agnew, 1987: 62).

The key element in American penetration of Asia was the seizure of the Philippines in 1898 following the Spanish–American war. The USA is frequently presented as a reluctant coloniser of these islands, their acquisition being an unforeseen consequence of the war with Spain. However, plans for the occupation were laid before the war and the importance of the conquest of the Philippines for American interests in Asia and the Pacific clearly laid out (Kalb and Abel, 1971: 30–1; Lafeber, 1963: 410–11).

For most of the period of Spanish rule the Philippines remained little integrated into South East Asia trade. Manila's function as the pivot of the trade between China and Mexico resulted in the islands being more closely tied to South America than to Asia. From the 1780s the production of crops for export to Spain was encouraged. The expansion of, in particular, sugar and tobacco exports made the islands virtually self-financing during the course of the nineteenth century (Hall, 1981: 756). However, until the 1830s the Philippines remained largely isolated from the international trading system with trade and production tightly controlled by Spanish mercantilism. In these respects the islands strongly reflected the weakness of Spanish capitalism.

The loss of the South American colonies during the 1820s further increased the islands' isolation and vulnerability. Spain responded to this in 1834 by opening Manila to international trade. While this stimulated agricultural production, particularly of tobacco, it also opened the islands to the activities of British and Chinese traders. These largely supplanted the Spanish so that by the 1870s the islands were referred to as 'an Anglo-Chinese colony flying the Spanish flag'.[44]

The development of trade and production after 1834 resulted in the rapid growth of the middle class, comprising Chinese, Filipino and Spanish elements. By the 1870s this had formed the basis of a nationalist movement which in the last quarter of the nineteenth century was to increasingly challenge Spanish rule. This culminated in the revolutionary movement of 1896–8, and the declaration of the Republic of the Philippines.

The republicans initially greeted the American defeat of Spain with enthusiasm. However, it rapidly became apparent that the USA had no serious intention of recognising the independence of the Philippines. In December 1898 the United States and Spain signed a peace treaty under which America obtained the Philippines. The republicans responded by declaring war on the USA. This conflict lasted until 1904. During it the USA deployed over 100,000 troops and an estimated 36,000 Filipinos perished (Kalb and Abel, 1971: 33–4; Tate, 1971: 375). Thus, American colonial penetration of South East Asia was no less violent than those of Britain, France and Holland.

American annexation resulted in rapid expansion of the economy as a supplier of cash crops. By 1938 80 per cent of the Philippines' overseas trade was directly with the USA, a proportion that suggests greater economic dependence than was the case in the region's other colonial territories. Thus the Philippines had a unique experience of 300 years of Spanish rule under which elements of mercantilism survived into the mid-nineteenth century, a brief period of 'free-trade' incorporation and half a century of rule by what was to become the most advanced capitalist economy.[45]

3.3 Forms of dependency

The patterns of annexation and incorporation that operated in South East Asia show great variation. Not only did the various Western powers adopt different approaches, the reactions of the indigenous states were also extremely varied. The colonial powers had very different attitudes towards their respective colonial territories. In many ways these reflected the particular circumstances of the individual metropolitan states. The French and Dutch imposed very direct rule which was regarded as permanent – the colonial territories were seen as an extension of the home country – thus Indochina was 'la France d'Outremer' and the NEI 'Tropical Holland'. France, and more especially Holland, were more dependent on their Asian colonies than was the case with Britain. The degree of this 'dependency relationship' reflected the relative strength of British, Dutch and French capitalism as well as the extent and variety of their colonial possessions. However, in terms of the exploitation of the respective territories these were differences in degree rather than kind. While each interaction set an individual stamp on the states concerned it is important not to become preoccupied by these. As is argued in the next chapter, the economic and political forms that resulted had considerable similarity.

4

Uneven development: The establishment of capitalist production

4.1 Introduction

The territory acquired by the European powers before the nineteenth century retained the long-established relations of production. During the course of the 1800s and, most particularly, after 1870, capitalist relations of production were increasingly exported to South East Asia. Consolidation of control over the areas where these relations were being developed was both a necessary and integral part of the process. As we have seen in the case of Thailand, 'control' did not necessarily imply annexation. The crucial elements were the reserving and opening up of areas to the activities of metropolitan capital.

4.2 The growth of trade and production

The broadly based pattern of trade and production that existed prior to Western annexation of South East Asia underwent rapid and profound change. In general exports became narrowly based and oriented almost exclusively towards the needs of the individual colonial powers. The trade in high-value scarce products such as spice was supplemented, and later eclipsed, by the mass production of lower-value products. Many of these were long established in the region, such as teak, tin, coconuts, fibres and rice. Others were new: oil, rubber and palm oil. This process was initiated by the Dutch who began the large-scale export of sugar, coffee and later indigo from Java during the seventeenth century. These products rapidly supplanted spice in the VOC trade (Tate, 1979: 36). In the territory controlled by the Dutch the relations of production remained largely unchanged until well into the nineteenth century, the Dutch 'merely superimposing themselves upon an existing system of

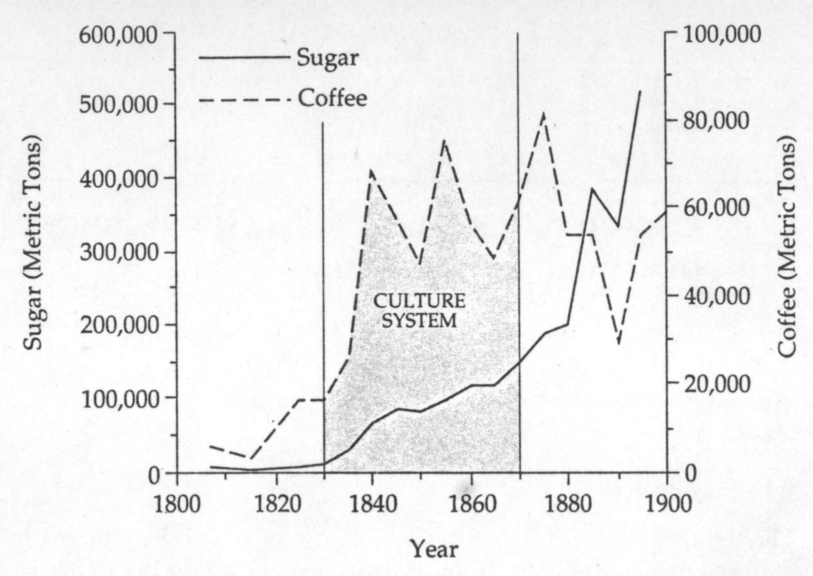

Figure 4.1 Growth of coffee and sugar production in the Netherlands East Indies, 1807–1900
Source: J. S. Furnivall, 1944: 75, 109, 129, 171, 127

peasant production in which surplus was extracted by means of political coercion' (Robison, 1986: 5). The system of expropriation was expanded and greatly intensified, reaching its height in the period 1830 to 1870 under the 'culture system'.[1]

The core of the system introduced by Governor General van der Bosch was:

> the remission of the peasant land taxes in favour of his undertaking to cultivate government-owned export crops on one-fifth of his fields or, alternatively, to work sixty-six days of his year on government-owned estates or other projects. (Geertz, 1963: 52–3)

Under the culture system there was a very rapid expansion of export crops, particularly coffee and sugar (figure 4.1). Most accounts suggest that the system virtually converted Java into one vast state plantation. However, the imposition was by no means uniform or complete; only 6 per cent of the cultivated area and 25 per cent of the population were directly affected in any one year (Geertz, 1963: 53).[2] The system operated most effectively in the sugar-growing areas; outside of Java it made only limited headway (Schrieké, 1929: 41). Activity in the outer islands was largely confined to tin mining in Banka and coffee production in Minahasa and, more significantly, in the 'culture zone' of West Sumatra (Chin Yoon Fong, 1977: 131).

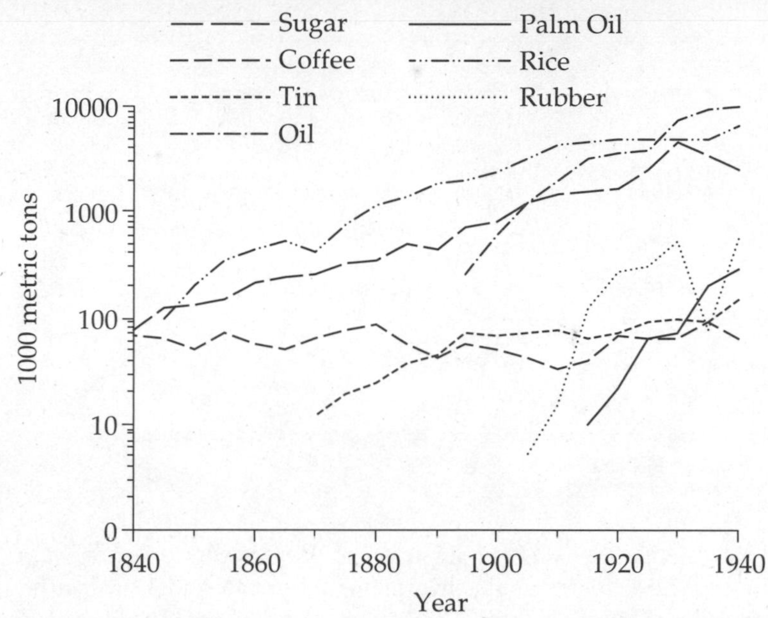

Figure 4.2 Growth of South East Asian primary exports, 1840–1940
Source: Tate, 1979: 572–578
Ingram, 1971: 38, 95
Murray, 1980: 180

Not until the 1850s was a beginning made in transforming this essentially mercantilist situation. Dutch capitalism was weak and, following the loss of the Belgium provinces in 1830, Holland was left with limited manufacturing capacity (Geertz, 1963: 47–8). Thus, unlike Britain, there was no imperative to open up markets for manufactured goods. Rather, the Dutch performed an intermediary role in the world-economy, supplying tropical raw materials and foodstuffs to the markets of the rapidly industrialising nations.

Outside of Java, European investment and involvement with production remained limited until the last quarter of the nineteenth century during which the expansions of South East Asian primary production accelerated (figure 4.2). This was directly related to the quickening pace of European industrialisation and the consolidation of political control over the region. From 1869 the opening of the Suez Canal and the rapid adoption of steam shipping facilitated the opening of South East Asia to the trade and production of the rival European powers.

The changes in the composition and control of South East Asian trade that occurred during the latter part of the nineteenth century are particularly well illustrated by Thailand. Here, following the Bowring

Table 4.1 *Thai exports c.1850*

Commodity	Baht	Commodity	Baht
Bark	110,000	Rice	150,000
Birds' nests	172,800	Pepper	99,000
Cardamons	124,000	Tobacco	100,000
Raw cotton	450,000	Tin and tin utensils	253,500
Cotton cushions and mattresses	211,500	Sticklac	254,000
		Sugar	708,000
Fish	213,500	Lard and fat	146,000
Iron and ironware	180,000	Sapanwood	350,000
Dried meat	120,500	Agilawood	100,000
Oil	101,000	Other items[1]	1,127,000
Hides, horns and skins	503,000		
Ivory	80,000	Total exports	5,585,500
Gamboge	31,000		

[1]. includes textiles and pottery
Source: Ingram (1971: 22 and 25)

Treaty of 1855, there was a rapid influx of Western enterprise. Between 1856 and 1958, for example, five major American and British firms entered the rice trade (Holm, 1977). Foreign trade, which before 1855 was almost exclusively in the hands of the Chinese, operating within the protective framework of the Royal Trade Monopoly, passed into the hands of Western, mainly British firms. By 1890 70 per cent of the foreign trade (by value) was handled in this way (Holm, 1977).

In 1850 Thailand exported a wide range of products (table 4.1). With the exception of rice all of these declined rapidly in importance between 1855 and 1870. Textiles and sugar processing, the largest indigenous manufacture, and both important sources of export earnings before 1855, were replaced by imports (Resnick, 1970; Chatthip Nartsupha and Suthy Prasartset, 1981: 2–6). Between 1870 and 1913 the value of cotton textile imports, mainly from Britain, increased sevenfold (Caldwell, 1978: 8). Land-use changes followed, with a decline in the area planted to sugar and cotton, particularly in the Central Plain, and some land being replanted to wet rice (Ingram, 1971: 36). Shipbuilding, the most highly developed manufacture in terms of organisation, finance and technology, declined sharply after 1855 (Sarasin Viraphol, 1978: 180).

By the late nineteenth century Thailand had become almost entirely dependent on rice exports (Feeney, 1979: 133, table 4.2). These were expanded to feed mainly immigrant plantation workers elsewhere in Asia. 'In this way Siam helped the imperialist powers in their well-organised design to extract more surplus value from their colonies'

Table 4.2 *Thai exports 1867–1950*

Period	% rice	% rubber[1]	% tin	% tin	% of all four
1867	41.1	0.0	15.6		56.7
1890	69.7	0.0	11.1	5.5	86.4
1903	71.3	n/a	6.4	10.4	88.2
1906	69.1	n/a	11.0	11.2	91.3
1909–10	77.6	n/a	7.8	6.4	91.9
1915–16	70.1	n/a	15.9	3.9	89.9
1920–24	68.2	0.8	8.6	4.5	82.1
1925–29	68.9	2.3	9.0	3.7	83.9
1930–34	65.4	2.0	13.8	3.9	85.1
1935–39	53.5	12.9	18.6	4.2	89.2
1940–44	60.5	12.1	11.6	1.6	85.9
1947	35.3	7.0	11.8	5.1	59.2
1948	50.5	13.4	5.9	3.4	73.2
1949	62.7	8.3	5.3	3.7	80.0
1950	50.8	21.8	6.7	3.8	83.1

1 Rubber production began in Southern Thailand during the early 1900s. The partial data on production and export in the period before 1920 is complicated by the loss of the main rubber-growing areas to Britain in 1909.
Source: Ingram, 1971: 94

(Chatthip Nartsupha and Suthy Prasartset, 1981: 393). From the early years of the twentieth century the expansion of tin and rubber production reduced this dependence. However, until the 1950s over 80 per cent of export earnings were derived from rice, rubber, teak and tin.

The development of mainland rice exports

During the second half of the nineteenth century the deltas of the Chao Phraya, Irrawaddy and Mekong were developed as major producers of rice for the international market. While the end result of these developments were similar the processes that operated showed considerable variation.

In Thailand the rapid expansion of rice resulted from the production of increasingly large surpluses by the established indigenous rice growers. Some of the expansion of production was initiated by wealthy landowners who purchased large areas of unused land, particularly on the east bank of the Chao Phraya river, the cultivation being carried out by tenant farmers (Mougne, 1982: 253). However, the expansion of production was predominantly the result of the activities of small-scale Thai farmers. The rapidity of this response to the process of incorporation can be interpreted as reflecting the extent to which the market economy was established in Thailand before 1855 (see 2.7). However, the mechanism which induced large numbers of farmers to engage in the arduous task of land clearance and land levelling is far from certain and merits considerably more research. A very limited role was played by the state of large-scale private undertakings. This situation contrasts strongly with the development of rice exports in Irrawaddy and Mekong deltas.

In Indo-China:

> The fundamental objective of colonial land policies . . . was to provide economic incentives for large-scale metropolitan concessionary companies and to attract a settler population in sufficient numbers to justify the tremendous financial burden of French occupation and administration. Colonial land policies date from the initial territorial conquest, when all lands that were not considered private property under French law were declared state domain. French settlers and private enterprises were permitted to claim the rights to uncultivated lands under a system of state-supervised concessions designed to encourage European immigration and proprietorship. The concession system applied only to French citizens and Vietnamese who had satisfied the requirements of assimilé status. The fundamental colonial distinction between the indigenous and non-indigenous populations became the foundation for two distinctly different systems of property rights. (Murray, 1980: 55)

Large-scale expropriation of land took place, particularly in the sparsely populated areas of Cochinchina. Between 1880 and 1904 543,393 hectares of land were expropriated and the area under rice expanded from 615,887 hectares in 1881 to 10,212,081 in 1900 (Osborne, 1969: 26). Very large land grants were made to French citizens and collaborative Vietnamese, initially in Cochinchina, but later in Tonkin and Annam (Dumont, 1935; Marr, 1971). A series of 'land rushes' took place in the 1890s and early 1900s, creating almost a 'latifundia' pattern with estates in some cases exceeding 30,000 hectares. By 1931 European concessions accounted for 10 per cent of the cultivated area in Tonkin, 17 per cent in Annam and 25 per cent in Cochinchina (Chesneaux, 1966: 168).

The rapid growth of rice exports from Cochinchina was closely related to land colonisation involving large-scale canal- and dyke-building operations. By the 1930s 4,000 kilometres of canal had been constructed and nearly 2 million hectares of marshland reclaimed (Murray, 1980: 179). During the 1920s and 1930s the colonial administration also developed the existing dyke and irrigation systems to increase and stabilise the production of rice in Tonkin and Annam (Murray, 1980: 182–4). The most rapid expansion of rice cultivation and exports took place in Burma (figure 4.3). Following the British annexation of Lower Burma in 1852 the Irrawaddy delta lands were opened up

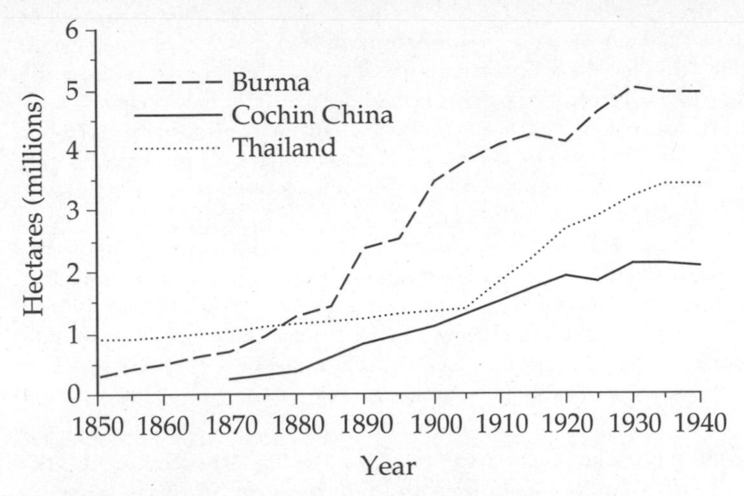

Figure 4.3 Expansion of rice cultivation, 1850–1940
Source: Adas, 1974: 60
Murray, 1980: 180
Ingram, 1971: 44
Tate, 1979: 571
Feeny, 1982: 23

for export production. Until the 1850s the delta lands were sparsely settled, producing a limited amount of rice for sale in what became Upper Burma. However, the delta was by no means the unsettled wilderness that some writers have suggested (Furnivall, 1938: 41; Hall, 1981: 746). During the preceding hundred years migrants from the densely populated Central Dry Belt had been slowly colonising the difficult but potentially highly productive lower delta lands. This process was greatly accelerated under the conditions created by the British annexation.

It is likely that the ban on rice exports, the low level of prices resulting from official control of the domestic market and the limited availability of goods for purchase constrained the development of commercial rice production (Adas, 1974: 23). While, as in Thailand, the relations of production appear to have developed to the point that once Burma was opened to the world market the peasant producers responded rapidly, the same reservations apply. Prior to 1852 the delta lands were operated by what were effectively small, 'owner-operators'. There is no evidence of the existence of a 'landed' class (Adas, 1974: 33–4). The British encouraged the continuation of this pattern:

The government strove to make rights to land easily attainable and at the same

> time to ensure that holdings came to be controlled by agriculturists and not by speculators and moneylenders. Several forms of land tenure were experimented with, but two types were dominant in the decades of rapid expansion. British officials initially favored the patta system under which the government granted tenure before the cultivator cleared his land. Grants were made by local thugyis, or district officers, depending on the size of the holding. Only persons who were able to prove that they were *bona fide* agriculturists could obtain patta grants, which averaged fifteen to twenty acres.
>
> The amount of land brought under cultivation by means of the patta arrangement was small compared to the amount claimed by virtue of 'squatter' tenure. Most cultivators, following pre-British precedents, became landholders by the act of clearing and cultivating a patch of jungle or scrub. After 1876 the government recognised the squatters' tenurial rights once they had occupied and paid revenue on their holding for twelve years. (Adas, 1974: 35–6)

Colonial revenues were used to develop the infrastructure of the delta. Roads, canals, embankments and railways were constructed, which facilitated the rapid growth of settlement and commercial production. By 1930 Burma had become the world's largest exporter of rice, a position that was held until 1941.

The expansion of rice cultivation took place with little development of technology or, initially, of the relations of production. However, from the 1890s Burmese cultivators became increasingly reliant on foreign capital, both British and Indian. This entered the Burmese economy largely through the Chettier money lenders (Johnson, 1981: 122–123). The dwindling number of migrants from the Dry Belt resulted in the expansion of cultivation in the delta becoming increasingly dependent on the influx of Indian labour (Johnson, 1981: 123).

The establishment of the 'rice bowls' of the Mekong and Irrawaddy deltas necessitated investment and the active involvement of the colonial state. In both cases the development of export-oriented rice production was concentrated in previously thinly settled areas. The contrast with Thailand is striking. Not only was state involvement and infrastructural development minimal but production spread out widely from an already comparatively densely settled area becoming in the process increasingly extensive in character. The area planted to rice increased from an estimated 0.97 million hectares in 1850 to 1.53 million in 1905 (Ingram, 1971: 44). Between 1850 and 1900 the main expansion in rice area took place in the Central Plain. In 1905–6 this area probably accounted for 98 per cent of exports. As well as its proximity to Bankok, the Central Plain had a network of watercourses and canals that provided a cheap means of transport for bulky commodities. A variety of observers (van der Heide, 1906; Grahame, 1924; Smyth, 1898) commented on the commercial cultivation of the Central Plain and the contrasting subsistence agriculture of the North and North East.

Within the Central Plain the expansion of production same from an extension of the cultivated area, the replacement of other crops, notably sugar and cotton, and a general increase in the intensity of production as transplanting replaced broadcasting in many areas (Johnson, 1975: 210). The development of commercial cultivation was accompanied by the development of a land market and wage labour. Land prices rose sharply and workers began to be hired on a seasonal basis not only from the Central Plain but also from the North East (Johnson, 1975: 229–34). In contrast to Burma there was no inflow of foreign labour or capital into the agricultural sector.

Expansion of production in the Central Plain brought unreliable and less fertile land into cultivation. Production could only be substantially and reliably increased by the development of large-scale irrigation and flood-control schemes. The need for these was reinforced by a series of crop failures between 1905 and 1912. Official recognition of the problem resulted in the establishment of the Royal Irrigation Department and the engagement of a Dutch Irrigation adviser – J. H. van der Heide. His report, published in 1903, represented what Ingram (1971: 82) has described as 'a brilliant statement of the irrigation needs of Thailand and the solutions to them'. The proposal centred on a barrage across the Chao Phraya at Chainat; the cost was estimated at £2.5 million over 12 years. This scheme was rejected, as was a reduced scheme costed at £1.5 million. Van der Heide remained as Director of the Irrigation Department until 1909. He proposed a series of small-scale projects, all of which were rejected. In 1909 van der Heide resigned and in 1912 the Department of Irrigation was abolished.

The rejection of these proposals reflects both the precarious nature of Thai independence in this period and the power of the British financial adviser[3] who was firmly against expenditure on irrigation:

> It is, in my opinion, impossible to think of embarking ... upon the gigantic irrigation project lately submitted by the Ministry of Agriculture: I do not think that project can be thought of for very many years to come ... Before we can think of a great irrigation scheme we must provide funds for the strategic railways which are essential if the outlying Provinces are to be properly governed. Those railways must be constructed out of borrowed capital and I am altogether adverse to borrowing money for irrigation at present in addition to money for Railway Construction. Such a course would be rash in the extreme.[4]

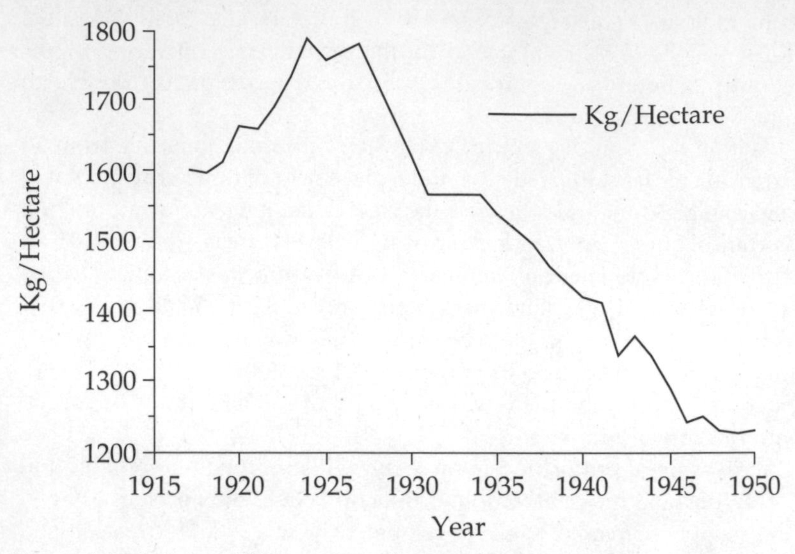

Figure 4.4 Thai rice yields, 1917–50 (five years moving averages)
Source: Thailand, Ministry of Agriculture (1972) *Agricultural Statistics*, Bangkok.

Brown (1978: 204), however, has argued that 'the Siamese ministers in the 1900s were essentially united in their decision to reject all proposals for large-scale irrigations works'.

In essence Brown explains policy in terms of how far it would contribute to the maintenance of the kingdom's political independence. Thus, priority was given to the maintenance of large currency reserves and 'to the construction of railways linking Bangkok with the more distant regions of Siam in order to secure the more effective government of the whole kingdom from the capital. The railways promised a clear strategical and administrative benefit. In comparison, though large-scale irrigation schemes would have undoubtedly encouraged the general economic development of the kingdom, it could not be said that they would have contributed directly to the maintenance of Siamese independence' (Brown, 1978: 211).

Whatever the interrelationships between the financial adviser and the Thai cabinets the policies of the early 1900s served the interests of British capital. Control of the outer provinces secured British interests, most notably in the Northern teak trade. The failure to develop irrigation and flood control in the Central Plain resulted in a rapid spread of commercial rice production into the less fertile and far more unreliable areas of the North and North East. By 1935 these areas were supplying nearly 25 per cent of exports. Production became steadily more extensive in character with average yields falling after 1920 (Feeney, 1979: 135–139; figure 4.4).

Tin

The development of tin mining during the nineteenth century stands in marked contrast to the expansion of rice production. The industry had been long established in the region, and from an early date the province of immigrant Chinese miners and traders.[5] Rapid expansion of European demand from the early 1800s stimulated the search for new fields and a rapid expansion of production particularly in the Malaysian Peninsula.[6] By the 1860s the Malay Straits had eclipsed NEI and Cornish sources (Tate, 1979: 186).

British traders were dealing in 'Straits Tin' during the eighteenth century, but not until the 1850s was any attempt made to become directly involved in production. These came to nothing. However,

> The high prices of the 1870s precipitated a bout of new attempts, particularly in Sungai Ujong and Selangor, but once again these ventures became ensnared in local politics and had to be abandoned. In fact the primary reason for the failure of Western enterprise to make headway in the tin industry prior to 1874 lay in the anarchy prevalent in the Malay tin states themselves. This anarchy retarded all mining enterprise in the Peninsula and prevented the full flowering of the industry until the spread of British control created more favourable conditions. (Wong Lin Ken, 1965: 48–50)

The tin interests were part of the pressure for the 'forward movement' in the Malay States discussed in chapter 3.

Rapid expansion of production followed from the imposition of British control in 1874. By 1883 the Malay states were the world's largest producer of tin. However, during the nineteenth century despite the introduction of Western technology, notably the steam pump, and British control of the export trade, production and finance remained almost entirely in Chinese hands (Wong Lin Ken, 1965: 60–4). A similar pattern of Chinese-dominated tin production developed in Southern Thailand, centred on Phuket Island.

By the standards of Western mining the capital employed in the Chinese undertakings was extremely small. While much of the finance was accumulated by small-scale producers from a variety of sources an appreciable quantity was advanced by major traders (Wong Lin Ken, 1965: 63–4). The majority of capital appears to have originated in the Malay States and Straits Settlements. Little was raised or repatriated to China. However, as late as 1902 the largest mining concern in Negeri

Sembilan was operated from Canton (Wong Lin Ken, 1965: 64). During the late 1890s the larger amounts of capital needed in mining and its increasing scarcity due to the quickening pace of development led the Chinese miners into dependence on Indian money lenders. In part this stemmed from the reluctance of Western banks to advance money to Chinese (Wong Lin Ken, 1965: 218).

The initial acceleration of primary production after 1870 took place principally through control over existing systems of production. These included areas in which capitalist relations of production had already begun to emerge. In the mining and, to a lesser extent, timber and estate agriculture sectors, Chinese capital was establishing a pattern of small-scale capitalist production for the world market. In general European control of international trade left national and regional trade in the hands of Chinese, Indian and indigenous traders. In the last year of the nineteenth century, the preconditions were laid for the rapid expansion of European investment, direct control over production and the widespread establishment of capitalist relations of production.

4.3 The development of capitalist production

The establishment of capitalist production in South East Asia as in colonial territories in general centred on the production of food and raw materials. This process was closely related to the export of European capital and the establishment of European territorial control.

Between 1860 and 1914 the UK was the main source of capital exports. While after 1918 this role was increasingly filled by the USA, in Asia British capital remained dominant until the 1939–45 war. By 1913 perhaps 10 per cent of British overseas capital was invested in India, 1.2 per cent in China, and 1.3 per cent in South East Asia.[7] However, as Davenport-Hines (1989: 4) has emphasised, these figures refer to total British investment rather than private business investment. Thus the majority of British investment in India and China comprised railways and government stock. While it is virtually impossible to disaggregate public and private investment in South East Asia prior to the 1940s it is certain that the region was of much greater significance to British business interests than the generalised figures for capital flows suggest. By 1914 British investment in South East Asia probably amounted to £25 million in the Malay States, £16 million in Burma, £2 million in Thailand and £3–4 million in the NEI and Philippines (Falkus, 1989: 117; Callis, 1942). At this date French investment in Indo-China was *c.* £20 million[8] (Robequain, 1944: 161) and Dutch investment in the NEI *c.* £12.5 million (Lindblad, 1989: 13). However, the main period of growth for metropolitan investment in South East Asia was the 1920s. Whatever doubts may be raised over the importance of South East Asian investment for the Western powers the scale of activity was sufficient to lay the basis for capitalist production.

The nature of Western investment and business activity in South East Asia (and indeed in the colonial and semi-colonial countries as a whole) has received scant attention.[9] Until the 1880s investment was largely confined to import–export and agency activities. These concerns dealt in indigenously produced commodities, generally obtained through contact with local, principally Chinese, middlemen rather than directly from producers. The increasing volume of imported goods was similarly fed into the established distribution system. From the 1850s some firms became involved in limited processing activities, notably saw mills and rice milling. Thus Western enterprises, largely located in major ports and collection centres, remained generally peripheral to the region's internal economies (Falkus, 1989: 130–3). However, their control of external trade put them in a position of increasing power over the internal traders and producers.

During the 1880s not only did agency houses begin to diversify into direct control over production, but specialised primary producing companies began to appear, most notably in the tin sector. The agency houses remained of considerable importance, not only in trading but also increasingly in finance. They were to become a major channel for metropolitan funds that flowed into primary production.

Until the direct involvement of Western companies in plantation, mining, lumbering and processing of primary products these activities were largely in Chinese hands. It is impossible to gauge the amount of Chinese capital that flowed into South East Asian trading and production activities during the nineteenth century. However, it was sufficient to finance spectacular expansion of production of such commodities as tin. From the 1880s Chinese business activity became increasingly secondary to that of Western firms. However, much Western activity remained dependent on Chinese-controlled collection and distribution systems and in some areas Chinese capital continued to successfully compete with, and effectively supplement, Western activity.

The role of the colonial state

In all the South East Asian colonial territories, the development of capitalist production was heavily dependent on the state apparatus (Murray, 1980: 49). At the simplest level this involved the creation of the conditions, including the legal framework for the development of land

and labour markets and infrastructure, which would encourage individuals and companies to invest and engage in entrepreneurial activity. Swettenham outlined this policy in 1894:

in the administration of a Malay State, revenue and prosperity follow the liberal but prudently directed expenditure of public funds, especially when they are invested in high-class roads, in railways, telegraphs, waterworks, and everything likely to encourage trade and private enterprise ... The government cannot do the mining and the agriculture, but it can make it profitable for others to embark in such speculations by giving them every reasonable facility, and that we have tried to do. (cited Sadka, 1968: 339–40)[10]

The colonial state also frequently facilitated the penetration of the market economy into areas of previously largely subsistence activity by the creation of cash needs[11] through the imposition of taxation and the creation of state monopolies in such items as salt, tobacco and alcohol.[12] This is best documented in French Indo-China, but played a significant role throughout the region. The necessary administrative frameworks and basic infrastructure were largely financed by colonial revenues principally derived from taxation, excise and state monopolies. In the Malay States and Straits Settlements these comprised duties on opium imports, licence fees for gambling-houses and pawnshops and export duties on tin. In general the reinvestment of state revenues paid rapid dividends. By 1900 the total revenue of the Federated Malay States exceeded $15.5 million and the value of their total trade approached $100 million (Drake, 1979: 274). Direct European investment in infrastructure, notably railways and port facilities began in the 1870s in Java and, to a lesser extent, the Western Malay States and the Straits Settlements (Khoo Kay Kim, 1972: 95–100); elsewhere such developments did not take place until the late 1890s (Harrison, 1963: 206–18).

Incorporation and the rural sector

To a limited extent the subsistence element in the rural sector was protected by colonial regulations. This was most notable in the Malay rice-growing sector (Krataska, 1983) and in Java where the Agrarian Law of 1879 had the dual purpose of allowing expansion of European estate production while 'protecting' the subsistence sector (Harrison, 1963: 210). However, these communities were to become major sources of labour and later of food for the export sector. In the long term the 'protectionist measures' exacerbated spatial and communal divisions.

The impact of incorporation into the market economy on South East Asian rural communities has been aptly summarised by Scott:

The growth of the colonial state and the commercialization of agriculture complicated the subsistence security dilemma of the peasantry in at least five ways. First, it exposed an ever-widening sector of the peasantry to new market-based insecurities which increased the variability of their income above and beyond the traditional risk in yield fluctuation. Second, it operated to erode the protective, risk-sharing value of the village and kin-group for much of the peasantry. Third, it reduced or eliminated a variety of traditional subsistence 'safety-valves', or subsidiary occupations which had previously helped peasant families scrape through a year of poor food crops. Fourth, it allowed landholders, who had once assumed responsibility for some of the hazards of agriculture, not only to extract more from the peasantry in rents but also to collect a fixed charge on tenant income, thereby exposing the peasantry more fully to crop and market risks. Finally, the state itself was increasingly able to stabilize its tax revenue at the expense of the cultivating class. (Scott, 1976: 57)

These changes operated far from evenly. Their varying intensity reflected the nature and strength of the capitalism of the particular colonial power. In the NEI the subsistence insurance element, although at a steadily declining level, remained much stronger than was the case in the pioneer areas of the Irrawaddy and Mekong deltas. In these areas the peasant cultivators were in a much more vulnerable position. They had become dependent on the purchase of essential items and were faced with greater fixed demands on their production. These needs could only be met by the sale of rice at prices determined by the international market.

The growing concentration of land ownership in the most commercialised areas increased the exposure of the peasants to market forces but was also a direct consequence of that exposure. As was discussed in 4.2 French policy created a pattern of large holdings which was reinforced by the operation of the market. In the Irrawaddy delta concentration was a direct consequence of commercialisation. The process made least headway in Thailand, though even here there was a gradual growth of tenancy and landlessness in the Central Plain during the 1920s and 1930s (Hanks, 1972: 141).

In the Irrawaddy delta lands the closure of the rice frontiers, falling yields and the rising prices of land and necessities forced increasing numbers of owner-cultivators into debt. Tenancy, which had begun to spread during the 1890s, increased rapidly after 1900. The process whereby the landless could make the transition to landed which had prevailed during the initial period of colonisation (4.2) was not merely halted but actually put into reverse (Sansom, 1970: 18–25).

Farmers had become increasingly dependent on the purchase of goods and services which had previously been produced within the

Table 4.3 *Occupancy of land in Lower Burma, 1925–26 to 1937–38*

Fiscal year	Total occupied area in thousand acres	Tenanted area as percentage of total occupied area	Area held by non-agriculturalists as per cent of total	Area held by resident non-agriculturalists as per cent of total held by non-agriculturalists
1925–26	10,340	43	27	25
1928–29	10,654	44	29	25
1931–32	10,734	49	38	21
1934–35	10,926	57	46	18
1937–38	11,300	59	47	18

Source: Golay *et al.*, 1969: 221

household. Thus they were seriously affected by the 58 per cent rise in the price of 'commonly purchased items' between 1913 and 1925.[13] This situation was reinforced by the sharp fall in the price of rice between 1929 and 1931.[14] In 1930 less than 70 per cent of farmers were able to repay their debts after the sale of their crops (Adas, 1974: 141).

From the early 1900s the delta lands were transformed into an area of large estates, landless labourers and small rented units held on short and insecure tenure (Furnivall, 1938). Farmers had little incentive to invest in land improvement. This reinforced the tendency for yields to fall and undermined further the viability of small holdings. On the large estates rice was cultivated by large, specialised 'gangs' of Burmese or Indian labourers.[15] Increasing amounts of land became owned by non-agriculturalists (table 4.3). Foreclosure on debts was the principal factor in this increased concentration of holdings (Adas, 1974: 127).

Thus the increased opportunities that the opening of the delta lands gave to migrants during the late nineteenth century were eroded after 1900. From a situation where the majority of the population of the area benefited from incorporation increasingly a small group of land owners, merchants, millers and moneylenders gained while the majority of the population experienced deteriorating living conditions.

Adas has summarised the situation:

Declining yields and unfavorable market conditions combined with rising costs gradually deprived tenants of the margin of profit which had allowed many to move up the social and economic scale to the status of landholders. In the early decades of the twentieth century the movement of tenants was increasingly downward to the level of landless, agricultural labourers. Tenancy became a dead end for most agriculturists or a temporary respite in their fall from cultivator-owners to landless labourers, rather than the avenue of upward mobility it had been in the nineteenth century. By the 1920s the labourer–tenant–landowner progression that was such a dominant feature of rural Delta society in the first phase of agrarian development was a thing of the past in most areas. (Adas, 1974: 147)

This whole process was greatly accelerated after 1920 by the increasingly uncertain international market.

Plantations

The first large-scale development of directly controlled European production took place in Java. The transition to capitalist production controlled by Dutch investment was directly related to 'the growing strength of the Dutch industrial and financial bourgeoisie' (Robison, 1986: 6).[16] From 1870, following the passing of the Agrarian Law, the state-organised culture system was replaced by corporate plantations. In the production of sugar the plantation sectors' share rose from 9 per cent in 1870 to 97 per cent in 1890 (Tate, 1979: 55). In essence, as Geertz (1963: 83–84) has stressed, 'the Amsterdam and Rotterdam merchants ... did not create, as they later came to claim, the Netherland East Indies estate economy. They bought it – and rather cheaply considering the social costs of its production'. By 1930 over 9 per cent of Java's cultivated area was under estates (Furnivall, 1944: 312). The incursion of the Dutch corporations and banks also undermined the position of the Chinese who had been active in the estate sector before 1870 (Robison, 1986: 17).

The estates sector was dominated by sugar until the 1920s. However, gradual diversification took place, most notably with the introduction of rubber after 1900 (table 4.4). With the backing of major banks, the corporations not only diversified and spread production to the Outer Islands but invested heavily in irrigation, railways, port facilities and processing, creating in the process 'a comprehensive agro-industrial structure practically unmatched for complexity, efficiency and scale anywhere in the world' (Geertz, 1963: 85–6).[17] Lindblad (1989:13) estimates Dutch investment of *c*.200 million guilders in 1885, *c*.750 million in 1900 and *c*.1.5 billion in 1914. This Javanese lead was generally followed elsewhere in the region from the 1890s onwards, most notably in the Malay States.

In the later part of the nineteenth century the expansion of commercial agriculture in the Malay States centred on pepper, tapioca, sugar and coffee. Only in the latter two was Western control significant. Elsewhere, as with tin, production was almost entirely in Chinese

Table 4.4(a) *Netherlands East Indies, principal plantation crops in 1930*

	Number of estates	Area planted (ha. 000)		Number of estates	Area planted (ha. 000)		Number of estates	Area planted (ha. 000)
Sugar	179	198.0	Rubber	1112	573.0	Tobacco	117	52.6
Coffee	431	130.3	Gutta-Percha	3	1.3	Kapok	140	17.6
Tea	323	126.9	Ficus	14	0.6	Cantala	13	6.8
Cocoa	21	5.3	Oil palms	48	61.2	Sisal	20	12.5
Nutmeg	26	2.1	Essential oils	99	10.8	Hemp	27	20.3
Pepper	28	1.7	Cinchona	125	19.0	Cutch	15	1.9
Cassava	8	3.3	Copra	698	51.6	Cassava	8	3.3

Source: Furnival, 1944: 319

Table 4.4(b) *Netherland East Indies, the contribution of estates to total production and export earnings in 1930*

% of production from estates	% of export earnings	
Rubber	52	22.6
Coffee	42	2.2
Tea	82	8.2
Sugar	99	6.5
Tobacco	70	5.6
Palm oil	100	2.4
Pepper	1	1.2
Copra	5	5.6

Source: Pelzer, 1963: 120

hands. The coffee boom of 1890–5 brought a rapid expansion of European estate production in Negeri Sembilan, Perak and Selangor.[18] While coffee production was shortlived, declining rapidly after 1903, it payed a vital role in laying the foundation of the European plantation sector (Jackson, 1968: 172).

The rapid expansion of rubber production after 1903 was by far the most spectacular development of European production in the Malay States. The high price of rubber during the first decade of the twentieth century, and the comparatively low cost of land, made plantation investment extremely profitable (Allen and Donithorne, 1957: 11). To a limited extent the colonial administration gave encouragement to the expansion of rubber (Jackson, 1968: 234–41).

The capital necessary to open up new estates or convert established ones to coffee was generally beyond the scope of the individual planter. Thus the rubber boom was closely associated with the establishment of the corporate plantation system. By 1908 virtually all the estates in the Western Malay States were in the hands of companies, most of which were registered in London (Voon Phin-Keong, 1976: 40–62; Jackson, 1968: 242–4). Between 1908 and 1910 almost £8 million of authorised capital was issued for Malay rubber plantations (Platt, 1986: 115). A key role was played in this development by the established agency houses which linked the planters with the source of capital in Europe (Drabble and Drake, 1981; Voon Phin-Keong, 1976: 62–7; Jackson, 1968: 245).

The expansion of the rubber sector, like that of other South East Asian primary production, was far from even, reflecting the fortunes of the international market. In general the succession of booms and slumps consolidated the position of the large European corporations. By 1930 rubber accounted for nearly 70 per cent of the cultivated area in the Peninsula and a similar proportion of export earnings. During the 1930s the development of oil-palm production diversified the plantation sector and further consolidated European control of large-scale production in the Peninsula (Tate, 1979: 215).

While the rubber sector was dominated by the large European companies, smaller-scale Chinese production remained important and small-holder production spread in the wake of the plantations making use of infrastructure and processing facilities.[19] Small-holdings were typically Malay-owned and comprised a few hectares cultivated as an adjunct to subsistence rice cultivation.[20]

The organisation of the Malay States' rubber industry contrasted strongly with that of southern Thailand. As with tin, Thai rubber production was effectively an extension of British activities in the Malay

States. The crop was introduced by Chinese migrants from the Malay States and production was dominated by small-holders (Landon, 1939: 70–3). While European involvement in production remained limited – there were only two European-owned plantations by the 1930s – almost all Thai rubber was exported to the Malay States and the Straits Settlements, much of it for further processing (Tate, 1979: 523; Stifel, 1973).

In contrast to Java and the Malay States the Indo-Chinese colonial economy remained heavily dominated by mercantile capital until the 1920s while tea, coffee, rice and sugar plantations were established in Tonkin, northern Annam and Cochinchina. It was the introduction of rubber that heralded the main expansion of the plantation sector and of French investment. The first rubber plantations were established in Cochinchina in 1912. However, they remained small in scale until 1924 as French metropolitan capital was reluctant to invest in rubber before this date. The 41 million francs invested before 1921 had been very largely raised in French Indo-China itself (Harrison, 1963: 213).

Between 1920 and 1927 the area under rubber in Cochinchina increased from 18,000 hectares to 54,000 hectares, and the number of plantation workers from 3,500 to more than 20,000 (Murray, 1980: 241). In general estates were located close to densely settled areas where the operations of the colonial economy were forcing the population into a perpetual search for wage labouring (Buttinger, 1958: 195). From 1918 much colonial legislation was aimed at easing and regulating the recruitment of labour.[21] These regulations were in part made necessary by the conflict of interest between the southern plantation owners and European entrepreneurs who were beginning to develop mining, manufacturing and craft production[22] in the Tonkin area.

One of the most distinctive developments of estate production took place in the Philippines where after 1898 the USA built on the *hacienda* system established by the Spanish. Until the 1850s the production of export crops, principally hemp, rice, sugar and tobacco was the province of small-scale producers. The processing and trade in these commodities, particularly sugar, were largely in Chinese hands. During the second half of the nineteenth century the opening of the Philippines to the international market stimulated the development of large-scale estate production. However, this was constrained by a shortage of capital, the weakness of Spain's international trading position and the system of land holding. Most estates were cultivated by large numbers of small tenants, many of whom were effectively debt-bonded. This situation was compounded by much of the best and most strategically located land remaining in the hands of the major religious orders. These barriers to the development of commercial agriculture were rapidly reduced under the American administration.

While much American land policy was ostensibly directed at securing the position of the small producer, in practice it generally facilitated their removal and the establishment of large commercial estates. This is illustrated by the handling of the 'friar lands' as the holdings of the religious orders were termed.

The persistence of the religious holdings were both a barrier to the further development of capitalist production and a focus of Filipino hostility. Thus 'it would avoid some very troublesome agrarian disturbances between the friars and their quondam tenants if the insular government could buy these large *haciendas* of the friars and sell them out in small holdings to the present tenants' (Report of the Philippines Commission, 1900, cited Grunder and Livezey, 1951: 127). The purchase of these lands necessitated protracted negotiations with the Vatican. In 1902 the majority of the friar lands were purchased by the American government for $7.3 million. The disposal of land to the tenants made little headway and from 1908 land was allowed to be purchased in large blocks by corporations (Grunder and Livezey, 1951: 130).

The USA had no real commitment to land-reform beyond the redistribution of the religious holdings. There were no proposals to purchase the private *haciendas* which dominated the agricultural sector (Salamanca, 1968: 152). The 1902 Land Act limited the scale of private lands to 16 hectares for individuals and 1,024 hectares for corporations. This did not apply to the former religious holdings and was reportedly on occasions circumnavigated (Fryer, 1979: 202).

The role of the large land-owners, particularly those producing sugar, was enhanced by the American policies. Sugar was given preferential treatment in the American market (Stauffer, 1985: 244) which by 1914 absorbed 75 per cent of production (Salamanca, 1968: 135). American capital moved into trade, processing and, to a lesser extent, estate ownership, rapidly squeezing out British interests (Stauffer, 1985: 244). By the 1930s the Philippines economy revolved around the estate production of a limited range of products – sugar, hemp, copra, coconut oil and tobacco – mainly for the American market.

Between 1909 and 1939 the cultivated area increased from 2.8 million hectares in 1909 to 6.6 million hectares in 1939; much of this was for export crops. To a degree, restrictions on the size of estates, the sizable American estate investment in Hawaii and Cuba, together with uncertainty of continued easy access of Philippines produce to the USA tended to discourage American investment in estates (Grunder and

Livezey, 1951: 104–21; 133–6; 215–18). In addition the estate sector encountered serious problems in recruiting labour. From 1902 there was a ban on the importation of Chinese labour[23] and American companies complained of the difficulty of securing sufficient Filipino labour, its high cost, poor quality and the inadequacy of the labour legislation which prevented employers from compelling workers to honour their contracts (Elliott, 1963: 61–2). Voon Phin-Keong (1977: 17), in examining the failure of the Filipino rubber sector, concluded that 'restrictions on land alienation and labour recruitment from abroad constituted almost insuperable obstacles to the establishment of rubber plantations'. Many American companies preferred to invest in production elsewhere in South East Asia where there were no such restrictions.

By the 1930s Western–owned estates were a major force in the production of South East Asian export crops. They spearheaded the development of capitalist production in parts of the NEI, Malay States and French Indo-China, but remained of limited importance elsewhere. In Burma there was little development of estate production and in Thailand estates remained predominantly Chinese-owned. In the Philippines, despite the growth of American investment during the 1920s, the majority of the large estates were owned by Filipinos.

From the First World War the dominance of plantations in the production of export crops was eroded by the growth of the small-holder sector. In the NEI the share of plantations fell from 90 per cent in 1914 to 60 per cent in 1938. By the latter year small-holders produced 60 per cent of rubber, 50 per cent of coffee, 25 per cent of tea and virtually all the copra (Grigg, 1974: 235). The expansion of small-holder rubber production was even faster in the Malay States. By 1917 their area was as great as that under plantations in many districts (Jackson, 1968: 258) and in 1938 almost 50 per cent of rubber was produced on small-holdings (Voon Phin-Keong, 1976). In contrast small-holder production made little headway in French Indo-China (Harrison, 1963: 213).

Thus by the late 1930s, while the estate sector remained a major element in South East Asian production, the majority of exports were coming from small-holders. The rise of farmers who produced for export reflects the degree to which capitalist relations of production had spread through the region.

Mining and forestry

The initial movement of European capital into the exploitation of South East Asian forestry and mineral resources is most clearly illustrated by the development of the mainland tin and teak sectors.

Teak, although widely used for construction and shipbuilding in mainland South East Asia, was only exported in small quantities until the involvement of Western traders during the nineteenth century. Until the 1850s rising world demand was largely met by the exploitation of the forests of Southern India. However, following the British annexation of Tenasserim in 1826, Moulmein developed as a major exporter of teak cut by Burmese foresters, some of whom also operated in Northern Thailand (Falkus, 1989: 133). By the 1860s and 1870s probably the majority of Moulmein's exports comprised Thai teak rafted down the Salween (Ramsey, 1976: 22–3). Following the colonisation of Lower Burma in 1852 British traders began to export teak cut in the Pegu area (Falkus, 1989: 133). During the 1850s rising world demand, particularly for shipbuilding and railway construction, and the exhaustion of the southern India forests, led British firms to seek forestry concessions in Upper Burma. Following the seizure of Upper Burma in 1886 the Burmese foresters were gradually eliminated from the industry.

The continued expansion of demand, the protracted British 'pacification' of Upper Burma, the removal of the most accessible reserves and the imposition of conservation measures in Lower Burma served to focus the attention of the logging concerns on the direct exploitation of the Thai forests. Here the cutting was in the hands of Lao and Chinese logging concerns, the latter controlling the export trade through Bangkok. From 1882 the Borneo Company had agents in northern Thailand and by 1885 this concern had become the largest logging operation in the kingdom. In 1902 four British companies were cutting 60–70 per cent of the teak (Falkus, 1989: 143). By this date the largest of these concerns, the Bombay Burmah trading company, had also established a virtual monopoly of teak supplies to Rangoon and Moulmein (Macaulay, 1934: 3–5).

The demand for timber for local construction and fuel as well as the growing demand for tropical hardwoods resulted in the development of colonial forestry industries throughout the region. In French Indo-China, for example, timber production for local consumption rose from 1 million cubic metres in 1911 to 2.2 million in 1936 (Maurand, 1938: 820). From the 1890s hardwood exports developed in the NEI, North Borneo and the Philippines. These sources gradually replaced Burma and Thailand as the major producer. By 1937 North Borneo was exporting 178,000 cubic metres of hardwood compared to Thailand's 85,000. The vast reserves of the outer islands of the NEI, however, were little exploited until the end of the colonial period, Dutch forestry activity being largely concentrated on the more accessible teak reserves of northern Java (Fryer, 1979: 333).

In the exploitation of South East Asian timber reserves the European companies experienced little competition from indigenous and Chinese concerns because of the limited scale of their activities. This contrasts strongly with the Malay tin industry where European companies initially found difficulty in effectively competing with the Chinese enterprises.

British investment in tin mining dated from 1882, but, until the early 1900s, few undertakings were successful. Indeed most of the forty-seven tin mining companies floated between 1882 and 1900 failed (van Helten and Jones, 1989: 164). Mining remained firmly in Chinese hands; only in smelting, where British firms had a marked technological superiority, was Chinese capital displaced. Until the exhaustion of the richest and most easily worked alluvial deposits Western mining concerns had no technological advantage over the labour-intensive Chinese methods (Hennart, 1986: 133). In addition, Chinese mines had much lower labour costs than Western undertakings utilising similar methods (Wong Lin Ken, 1964: 138–9). However, the introduction of the steam-dredge during the early 1900s enabled the working of low-grade deposits in swampy areas that could not be worked by Chinese methods. Few Chinese firms were able to invest in dredges and from this period Western firms obtained decisive cost advantages (van Helten and Jones, 1989: 164). Their share of production increased from 10–15 per cent in 1906 to 64 per cent in 1936 (Wong Lin Ken, 1965: 211–30).

After 1900 Western firms began to extend their control into the tin-mining areas of Southern Thailand. The Thai industry became increasingly an annexe of that of the Malay States. The Thai–Chinese smelting sector declined rapidly and by the 1930s all ore was processed in the Malay States. Similarly, from 1907 Western-owned dredging firms began to supplant the Chinese-controlled hydraulic production, though labour-intensive methods retained their importance. In 1950 there were 324 mines in operation, of which only 25 employed dredgers though these accounted for 64 per cent of production. Only three dredgers were Thai–Chinese owned and the rest were British and Australian (Ingram, 1971: 100).

Following the development of European tin mining the the Malay States, similar activities were established on Banka in the NEI and in Tonkin. The Banka deposits had long been exploited by Chinese miners but the industry was less well developed than in the Malay States and was effectively suppressed by the imposition of a Dutch monopoly.

While a wide range of other minerals was exploited in the pre-colonial period, including zinc (notable in Tonkin), oil (at Yenangyaung in Burma) gold (most significantly in Borneo), copper, iron and precious stones – there was little continuity with colonial developments. From the early 1900s these and other minerals[24] were extracted by almost exclusively Western, capital-intensive mining operations. By the 1930s while a number of minerals, notably oil from Burma and the NEI, were of major significance to the respective colonial economies in terms of international mineral production, South East Asia's importance continued to rest firmly on tin (table 4.5).

The role of the Chinese and Indian immigrant populations

A major element in the development of capitalist relations of production was the large-scale immigration of Chinese and Indian workers. The colonial economies gave rise to a social and economic environment which was particularly attractive to the inhabitants of the densely populated provinces of Southern China and India. As Unger (1944: 201) has stated:

> Western capital and management were available in most places but there was a lack of adaptable cheap labour to do the work of clerical and supervisory personnel, and of traders and merchants to set up services and provide the goods needed in remote or newly developed areas.

In many instances, the colonial powers saw the indigenous population of South East Asia as unsuited to employment in the 'modern sector' of the economy. Fisher (1971: 181) neatly summarised this view:

> with few exceptions, the familiar insouciance of the indigenous peoples rendered them unresponsive to the stimuli of a money economy and notably reluctant to abandon their established way of life in order to work long and regular hours which the Europeans regarded as essential for the successful running of their enterprises.

Low population densities over most of the region limited the possibility of transferring labour wholly out of the traditional sector. The indigenous population had little control over trade and production; enclaves of Chinese, Indian and Arab traders and entrepreneurs controlled the trade and much of the production. Influential colonial administrators,[25] especially the British in the Straits Settlements, held favourable views of the Chinese as workers and entrepreneurs. Except in the Philippines, there were few restrictions on immigration until the 1930s.[26]

Chinese migration into South East Asia began to expand rapidly during the late eighteenth and more especially the early nineteenth centuries. By 1810, for example, there were probably over 40,000 Chinese in the west Borneo goldfields (Jackson, 1970: 24).[27] The rate of migration increased as the nineteenth century progressed, Chinese flocking to the mines, plantations and ports of South East Asia. The

Table 4.5 *South East Asian primary production, 1937–40, in metric tons*

	British Malaya and Borneo	Burma	French Indo-China	Netherlands, East Indies	Philippines	Thailand	Total	% of world production
Abaca	1.2	–	–	–	183.0	–	184.2	95.6
Cinchona	–	–	–	10.4	–	–	10.4	80.0
Coffee	–	–	1.5	62.4	3.0	–	66.9	7.0
Copra/ coconut products	116.0	–	10.0	506.0	54.0	–	686.0	73.0
Kapok	–	–	–	20.0	–	–	20.0	70.0[1]
Maize	–	–	565.0	2037.0	427.0	7.0	3036.0	80.0
Palm oil	46.0	–	–	238.0	–	–	238.0	47.6
Pepper	–	–	–	20.0	–	–	20.0	70.0[1]
Petroleum	1000.0	1000.0	–	7400.0	–	–	9400.0	4.5
Rice	324.00	4940.0	3945.0	4007.0	2179.0	1771.0	17165.0	98.0[1]
Rubber	501.0	8.0	61.0	432.0	–	38.0	1040.0	85.2
Sugar	–	39.0	43.0	547.0	1076.0	19.0	1724.0	21.0[1]
Cassava	–	–	–	7759.0	–	–	7759.0	80.0
Tea	–	–	0.7	67.0	–	–	67.7	17.0[1]
Teak[2]	–	475.0	–	400.0	–	189.0	1064.0	95.0
Tin	77.0	2.0	1.6	40.0	–	13.4	134.0	65.0

[1] Percentage of world exports
[2] Cubic metres
Source: Tate, 1979: 25

Table 4.6 *Distribution of Chinese and Chinese investment in South East Asia during the late 1930s*

	Chinese %	Investment %
British Malaya (including Singapore)	39.0	31.4
British North Borneo and Sarawak	4.3	n/a
Burma	5.0	2.4[1]
French Indo-China	12.9	12.8
Netherlands East Indies	26.0	23.6
Thailand	11.0	14.1
Philippines	1.8	15.7
Total	100.0	100.0

[1] The Indian share of total investment in Burma in 1936 was 62.1 per cent
Source: Fisher, 1964: 181; Purcell, 1965

introduction of steam-shipping speeded and cheapened the process. In general, the migrants travelled under appalling conditions and their treatment on arrival was usually equally unfavourable; the excesses of this traffic led many to label it the 'pig trade'.[28]

Annual arrivals in Thailand rose from *c.*15,000 in the 1820s to a peak of over 35,000 in the early 1900s (Skinner, 1958: 59). In the Malay States, while Chinese penetration for mining and agriculture had assumed significant proportions before the direct intervention of the British in 1874, the main expansion took place after 1880. Between 1881 and 1900 nearly 2 million Chinese entered the Federated Malay State and over 1.5 million Perak and Selangor (Ooi Jin Bee, 1976: 116).

The great increase in Chinese immigration after 1800, although largely induced by the opportunities offered by the opening of South East Asia to the world market, was also stimulated by conditions in China. Southern China in particular was experiencing rapid population growth, a consequence of two centuries of stability under the Manchu dynasty, with limited expansion of food production. In the first half of the nineteenth century the deterioration of living conditions was aggravated by natural disasters and repeated outbreaks of political unrest (Tate, 1979: 21).

Until 1929 the Chinese population of South East Asia remained essentially transitory. After this date increasingly strict control of movements resulted in a greater permanency in the communities. Table 4.6 indicates the general distribution of Chinese investment in South East Asia at the end of the 1930s.[29] However, the general distribution of these groups conceals the very uneven distribution of immigrant populations within countries. In general the immigrants were heavily concentrated in urban areas. For example, by the late 1930s the Chinese represented 77 per cent of the population of Singapore, over 50 per cent of Bangkok, 30 per cent of Saigon-Cholon and perhaps 5–10 per cent of Manila and Jakarta (Fryer, 1979: 493).[30]

The majority of South East Asian Chinese remained poor, but a significant number came to control the internal trade and commerce of many of the region's economies. Southern China, the area of origin of the vast majority, had a long history of trade both internal, interregional and overseas. It is difficult to quantify the contribution made by the Chinese to the South East Asian economy; some indication of the scale of Chinese business activity in the late 1930s may be gleaned from figure 4.5. By this period outside of the NEI and Burma, Chinese investment in the region represented perhaps 38.7 per cent of total investment. In Burma the role of the Chinese in the economy was taken by Indian migrants, while in the NEI the scale of Dutch investment and the more restrictive economic policies of the colonial administration resulted in an over-shadowing of the Chinese.

In all the countries of the region other than Burma the Chinese came to dominate petty trading, money lending, small-scale production and the collecting and processing of much agricultural produce. It has been estimated that by the late 1930s, 80–90 per cent of the rice mills in Thailand, over 80 per cent of those in Indonesia and 75 per cent of those in the Philippines, were owned by Chinese (Fisher, 1971: 182).

In contrast to the widespread Chinese migrations into the region, the Indian presence after 1834 was mainly confined to the British territories and comprised a smaller number of people with generally less economic power (Kaul, 1982).[31] Most Indians arrived as labourers. The majority of the migrants into Burma and the Malay States were Tamils from Madras who generally failed to rise from the level of labourers because of their conditions of employment (Tate, 1979: 23). The average Indian labourer came as a recruit to work in specific plantations (Singh, 1982). From the 1880s this recruitment was the subject of increasing official supervision. Short-term contracts, controls over movement, and strict discipline and regimentation prevented these groups from establishing themselves permanently in the region. The Indian migrants, although less varied in language and culture than the Chinese, were by no means homogeneous. The dominance of Tamils should not be allowed to obscure the variety in caste, language and religions of this group. The Chettyars, among the best-known and documented of the smaller groups, originally a money-lending caste, came to dominate many parts of the Burmese economy (Mende, 1955: 118).

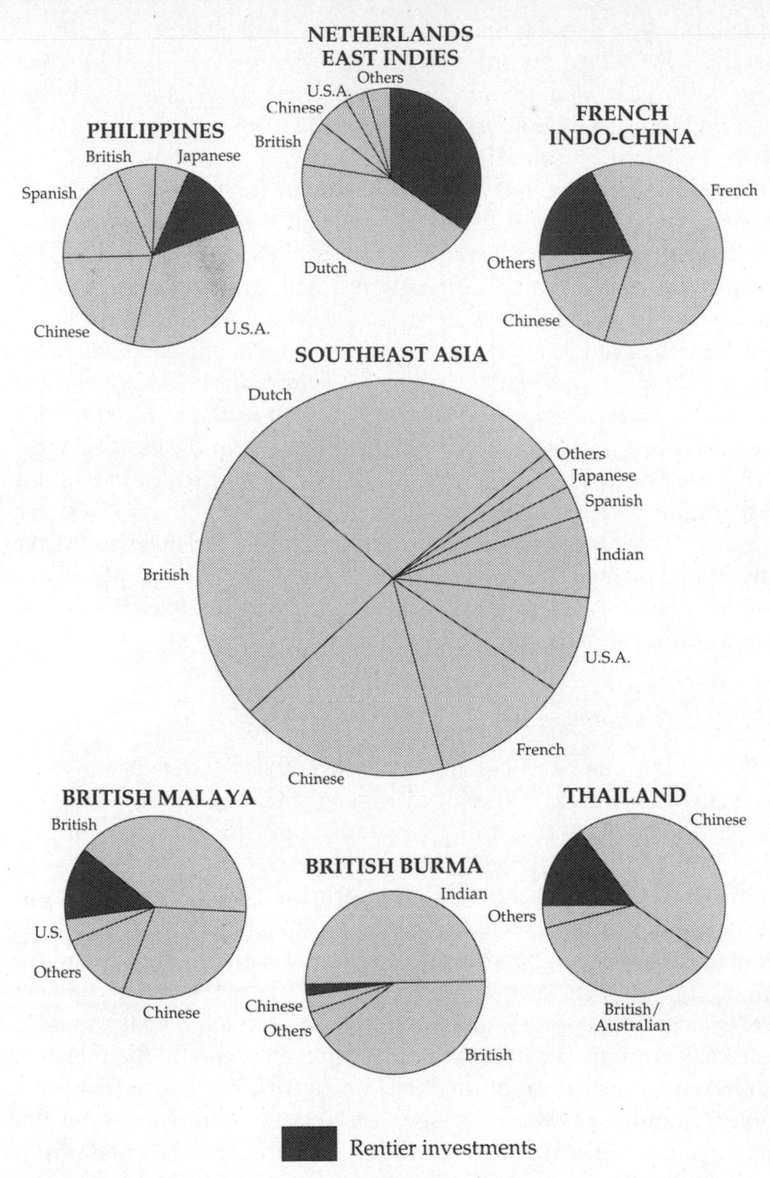

Figure 4.5 Capital investment in South East Asia during the late 1930s
Source: compiled from a variety of sources by Tate (1979: 28)

In addition to these labouring recruits a small number – educated clerks, teachers, medical orderlies – were brought in to serve the needs of the British colonial administration. Similarly, merchants and entrepreneurs began to migrate to take advantage of the new opportunities offered in Burma and the Straits Settlements.

By 1930 there were one million Indians in Burma. They held the largest share of investment, owned over half the cultivated land and dominated trade and commerce.[32] In the Straits Settlements the Indians were far less numerous, and apart from small groups of professionals they comprised mainly labourers descended from indentured recruits.

By the end of the colonial period the Indian and Chinese communities formed integral parts of the region's economy. In 1938 Chinese and Indian investment represented some 23 per cent of the regional total.[33] The immigrant population provided key elements in the establishment of capitalist production in South East Asia. They provided capital for local trading, which remained essential but unattractive to Western investors. They were the core of a working class and likewise of the petty-bourgeois and capitalist classes. In the post-colonial period the resulting plural societies and associated 'racial divisions of labour' were to be a source of major problems (chapter 5).

Industrial development

In South East Asia the colonial powers created extremely lopsided economies with highly developed mining and plantation sectors, a neglected indigenous sector and an industrial sector of minute proportions. Industrial development was initially concerned with the processing of primary products into a suitable form for export. Production was heavily concentrated in such activities as tin smelting, rice milling, sugar refining, rubber curing, extracting oil from oil palm and copra and the canning of products such as pineapples. While plant and equipment were principally imported, in all the economies a limited light-engineering sector developed to service, repair and increasingly manufacture simpler components. Some of these involved local, most frequently Chinese, capital; this was most apparent in the construction and maintenance of rice mills. Similarly the major port-cities gave rise to a variety of ship repair, engineering and associated activities.

Colonial tariff structures largely eliminated indigenous South East Asian manufacturing and severely limited the possibilities for the development of a modern manufacturing sector. However, a range of low-grade manufactures did develop to meet the needs of the local population. These activities involved both local and metropolitan capital. Such developments were additional to, and largely separate from, the domestic craft activities that survived throughout the region, particularly in the more densely peopled areas. In Tonkin in 1936, 250,000 people, or rather more than 7 per cent of the adult population, were engaged in craft activities for the greater part of the year (Gourou, 1936: 460–505.[34]

The development of textile production in French Indochina illustrates the constraints under which colonial manufacturing operated. Despite strong opposition from the French authorities to any form of industrialisation, numerous manufacturing industries emerged (Murray, 1980: 343). The distance from France and the availability of cheap labour led French firms to invest in the production of principally low-grade goods for the local market. Primary processing and the manufacture of building materials, spirits, cigarettes and pottery posed no serious threat to French imports. In contrast textiles were a major item of trade representing 26.6 per cent of Indo-Chinese imports in 1938. Mitchell (1942: 158) explains the development of textile production:

> When the French occupied Tongkin, they found that the native weavers were importing large quantities of cheap yarn from foreign countries, particularly British India, with which expensive French yarns could not compete. The only way in which France could secure for herself the profits offered by Tongkinese demand for yarn was to establish a modern spinning factory in Tongkin itself, where labour was abundant and cheap as in Bombay, and where it was hoped that locally grown cotton could provide the necessary raw materials.

The first French textile plant was established in Hanoi in 1894. By 1938 the industry employed *c.*10,000 people and produced 11,000 tons of yarn, most of which was sold to village hand weavers, as well as over 1 million blankets and over 3,000 tons of cotton cloth (Murray, 1980: 350). However, Indo-China was unable to produce sufficient raw cotton, 80–90 per cent of which was imported from India and the USA.

The merits of industrial development in the colonial possessions was the subject of official debate, most notably in the NEI. However, the prevailing tariff situation and the interests of metropolitan capital largely precluded development under normal trading conditions. the disruption of world trade during the 1914–18 war brought a brief period of expansion to South East Asian manufacturers both to supply local needs and to contribute to the 'war efforts'.[35] In the NEI, for example, an industrial section of the Department of Agriculture and Commerce and a technical institute were established. Such developments were seriously curtailed following the reestablishment of 'normal trading'.

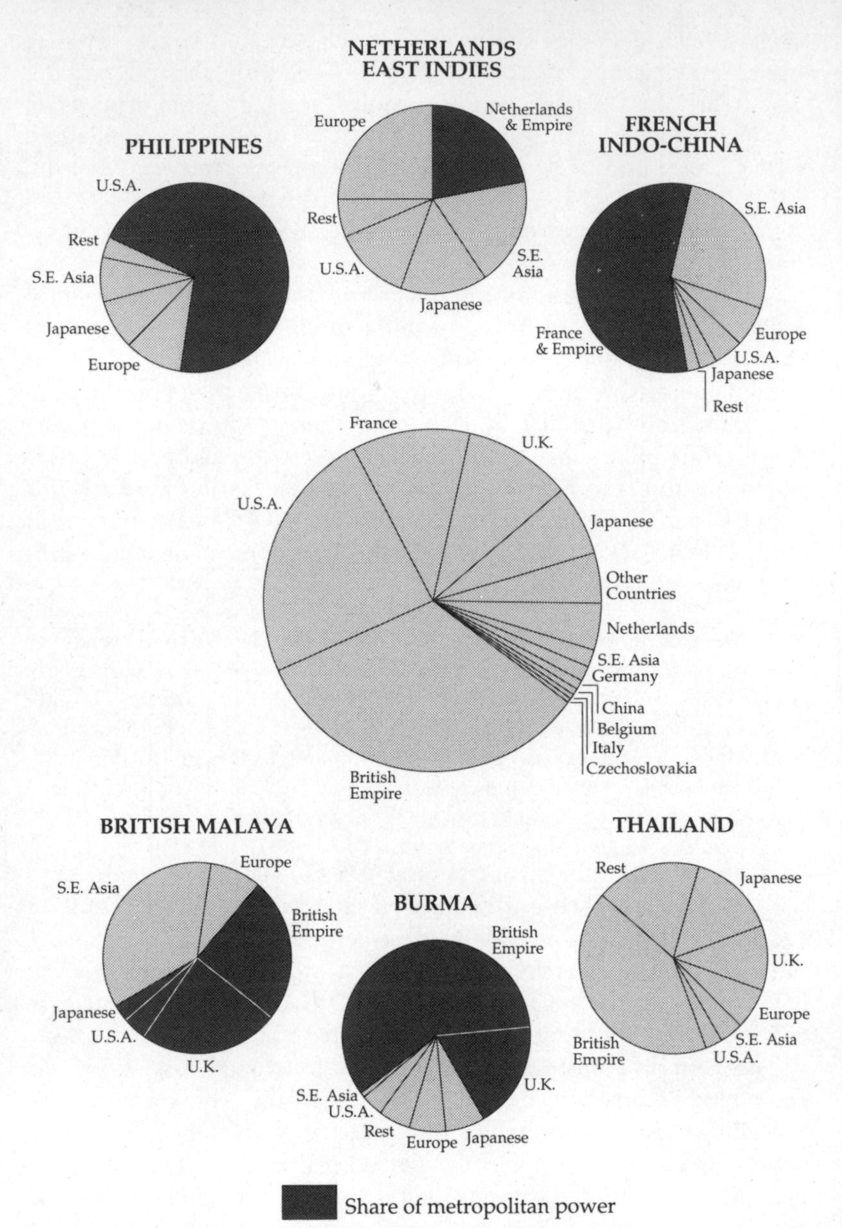

Figure 4.6(a) Origins of South East Asian imports, share of total value, 1938
Source: compiled from a variety of sources by Tate (1979: 25)

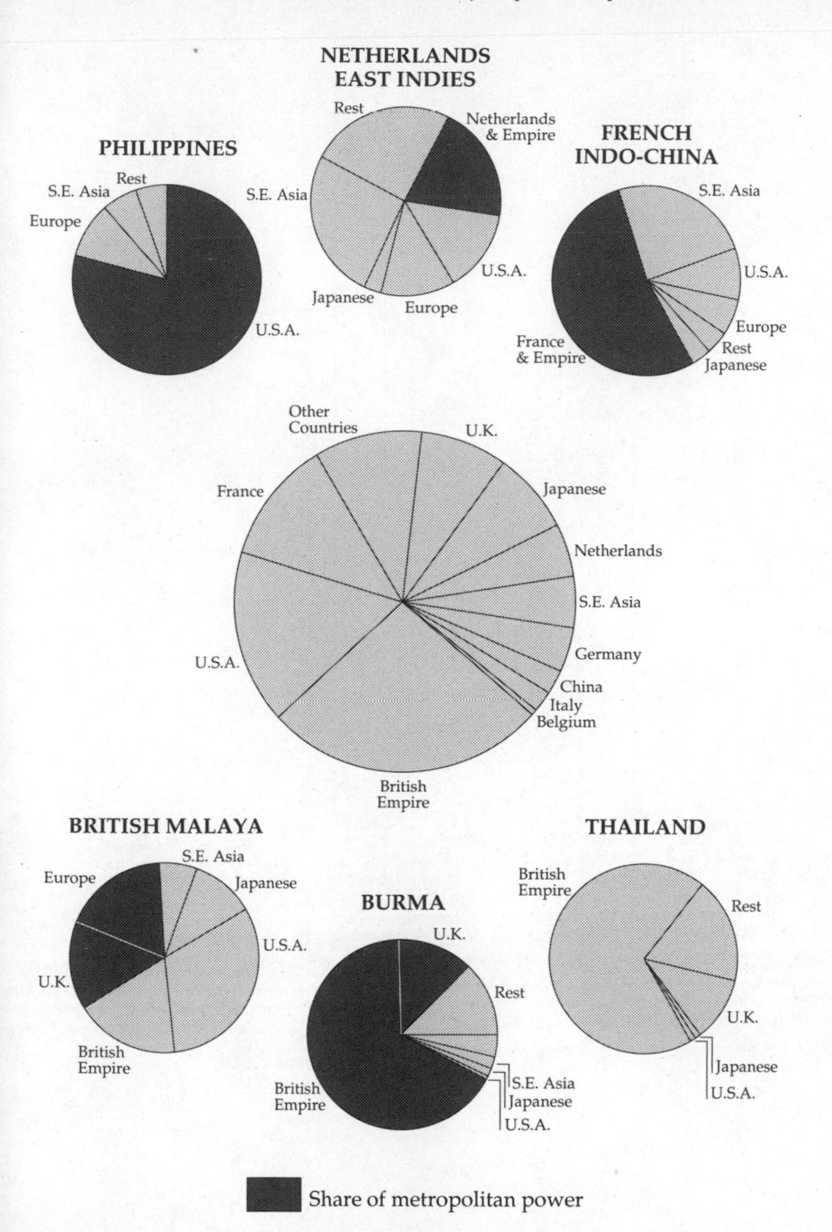

Figure 4.6(b) Destination of South East Asian exports, share of total value, 1938
Source: compiled from a variety of sources by Tate (1979: 25)

Table 4.7 *Distribution of French investment on the basis of new capital issues of Indochinese Corporations, 1924–30 and 1931–38*

	1924–30		1931–38	
	million francs	%	million francs	%
Agriculture	1,272.6	33.4	214.9	27.0
Mining	653.7	17.3	73.4	9.2
Industries (manufacturing) Utilities	606.2	15.9	168.5	21.2
Transportation	174.2	4.6	70.7	8.9
Commerce/trade	363.6	9.5	49.8	6.3
Banking and real estate	744.1	19.5	89.7	11.3
Total	3,814.4	100.0	796.4	100.0

Source: Callis, 1942: 78–9

The 1920s were by far the most important period of expansion for European investment in South East Asia. In particular the boom of 1924–30 saw sharp increases in investment and production throughout the region (Callis, 1942). This consolidated South East Asia's position as a major producer of raw materials (table 4.6). In terms of trade and investment although the various territories appeared to be firmly locked into their respective imperial economies (figures 4.5 and 4.6), during the inter-war period these structures began to weaken and signs of economic diversification emerged. While the majority of metropolitan capital continued to move into primary production, increasing amounts went into manufacturing and finance (Callis, 1942). During the 1930s there was not only a sharp fall in total investment but a marked shift from primary production to manufacturing (table 4.7).

The recession of the 1930s gave rise to a second and more sustained period of manufacturing development. In part the official encouragement given to manufacturing during the 1930s represented an attempt to appease nationalists' demands for industrialisation and to stimulate income growth. The establishment of tariff and infrastructure such that local capital could move into more profitable manufacturing activities was a common demand of sections of all the region's nationalist movements.

Some growth of manufacturing did result from increases in tariff protection and the imposition of import quotas, for example, in the NEI against Japanese textiles. However, the movement of capital into the industrial sector overall owed little to colonial policy. It was primarily a result of local and foreign capital exploiting the opportunities that the recession and the weakening of the established imperial structures offered. Much Chinese capital that had been closely tied to Western interests, often through 'agency houses', began to look for alternative outlets because of the way the recession was passed on. As a result capital moved into a variety of secondary activities and finance (Turnbull, 1982: 139). In Java between 1930 and 1937 the number of textile mills increased from 19 to 1,123. This was mainly the result of Chinese previously engaged as middlemen in petty commodity production moving into factory production (Robison, 1986: 24).

By the late 1930s a wide range of activities involving both local and foreign capital had been established (table 4.8). In terms of contribution to national income and employment the manufacturing sectors remained limited.[36] However, by 1940 the basis of a manufacturing sector and the core of an industrial working class had been established in all the region's major territories.[37] The organised industrial workers

Table 4.8 *Ownership of the manufacturing sector in the Netherlands East Indies during the late 1930s*

Commodity	No. of plants	Ownership
Beer	2	Western
Margarine	2	Western
Oxygen plants	4	Western
Rubber products	1	Western
European shoes	1	Western
Automobile assembly	1	Western
Light bulbs	4	Western
Dry-cell batteries	1	Western
Packaging material	1	Western
Paper	6	2 major plants, Western; ownership of 4 smaller plants not given
Paints	1	Major plant controlling 70% of production, Western; other plants not given
Cigars	3	1 Dutch; 2 Chinese
Glass	5	1 Dutch; 4 'Foreign Asiatics'
Soft drinks	130	Largest producer British; most of remainder Chinese
Soap	13	Largest producer Dutch; most of remainder Chinese
Western-type cigarettes	24	Leading producers: 1 Dutch, 1 Belgian, most of remainder Chinese
Stationery supplies	not known	Leading producer Dutch; remainder not known
Rice mills	1,137	Nearly all Chinese
Ice manufacturing	204	Leading producer Dutch; most of remainder Chinese
Dairies	not known	All Dutch
Engineering workshops	110	Nearly all Dutch
Sawmills, ice cream plants, perfumed cigarettes, modern furniture, repair and workshops	not known	Predominantly Chinese

Source: Sutter, 1959: 34–57, summarised and tabulated by Golay *et al.* 1969: 116

were to play a role out of all significance to their absolute numbers in the nationalist struggles of the 1930s and 1940s (see section 4.5).

4.4 The colonial legacy

Western control of South East Asia, when viewed in historical perspective, was of relatively brief duration, yet it brought profound social, economic and political change to the region. While development in pre-colonial South East Asia was far from even, the incorporation of the region into the world-economy during the colonial period radically altered the pattern of development, created deep divisions between territories and markedly uneven development at all levels.

The political divisions in present-day South East Asia are, with few exceptions, those agreed upon by the Europeans in the course of the nineteenth-century scramble for territory. This arbitrary parcelling out of the region into a patchwork of rival imperial structures has resulted in the post-colonial states 'facing the world with their backs to one another' (Boeke, 1944: 133).[38] Even adjacent territories of the same imperial power were poorly linked. It is particularly striking that no land communications were established between Burma and neighbouring Thailand, India and the Malay States. The low level of contemporary regional integration is a direct result of the 'watertight' imperial structures established during the late nineteenth and early twentieth century.[39]

Some of the political units created by the Western powers did correspond to earlier groupings, most notably Burma; the same cannot be said for the Philippines or the Indo-China states (Fisher, 1964: 160). However, even where there was some historical precedent the precise, permanent linear boundaries were essentially alien. Long-term problems have resulted, particularly with respect to the minority peoples of the mainland who formerly occupied the broad frontier regions separating the pre-colonial cores. In addition some boundaries were not clearly demarcated, resulting in protracted post-colonial disputes.

Population

Within the respective territories striking changes took place in the size, composition and distribution of population. While there appears to be a general relationship between these changes and the incorporation of the region into the world-economy the actual processes that operated are far from clear. It should be stressed that South East Asian population data prior to the twentieth century are far from reliable.

New crops, often associated with plantation development, were important in opening up previously thinly peopled areas, for example in the Annamite ranges of Indo-China (Buchanan, 1967: 85). In the Mekong, and Irrawaddy deltas and to a lesser extent in Thailand, large-scale movements of population resulted in the creation of a series of densely settled areas (Grigg, 1974: 103–5). Similarly the pattern of settlement was filled out in Java and the northern and central Philippines. Overall, the coastal fringes of the region took an increased share of the population (Fisher, 1964: 177). The 'negative' areas of the pre-colonial period remained largely untouched; the main impact of the colonial economies was confined to the already comparatively densely peopled areas. Increases in density from the early nineteenth century are in certain areas striking. In Java and Madura, population density increased from 96 persons per square kilometre in 1817 to 940 by 1940 (Gourou, 1958: 110). Some locally important increases in density come from mining developments after the mid-nineteenth century, most notably in the western Malay Peninsula, southern Thailand, north-western Burma and parts of Borneo. Overall the 'juxtaposition of densely occupied areas and sparsely peopled expanses' (Gourou, 1958: 106) remained largely undisturbed and possibly became more marked (Buchanan, 1967: 86).

For most parts of South East Asia available evidence points to an acceleration of population growth during the nineteenth century (see Myrdal, 1968: 1396–7; Fisher, 1964; Dixon, 1990b).[40] Much attention has focused on Java and the Philippines where the data are most extensive and probably most reliable. On the basis of a wide range of material Boomgard (1980: 45–7, cited Owen, 1987a: 46) suggests that in Java the rate of population growth increased from 1.4 per cent between 1755 and 1815, to 1.65 per cent between 1815 and 1850 and to 1.75 per cent between 1850 and 1900. Similarly for the Philippines Owen (1987a: 46) has calculated a rate of *c*.1.0 per cent for 1736 to 1800, 1.65 per cent for 1800 to 1876, which falls to 0.9 per cent 'in the series of mortality crises that characterised the quarter century before the census of 1903'. While the sparser and generally more dubious data available for the rest of the region are generally compatible with these trends, it is likely that their accelerations occurred later in the nineteenth century than those of Java and the Philippines (Fisher, 1964; Dixon, 1990b).

In Java from the early 1800s the Dutch and, between 1811 and 1816, the British improved agricultural techniques, particularly with respect to water control, drained swamps, and implemented elementary sanitation and hygiene measures. The Dutch were pioneers in tropical medicine, introducing smallpox innoculation and quinine (Furnivall, 1948: 362; Boomgard, 1987: 60–4) from the 1830s onwards. Forest clearance and swamp drainage further reduced the breeding grounds for mosquitoes and parasites (Wertheim, 1947: 92). The extent to which these developments were of sufficient magnitude to significantly reduce the death rate is open to question (Owen, 1987a: 51). Perhaps more importantly, in the Philippines such colonial health measures were largely absent.[41]

The degree to which the imposition of colonial rule brought stability and created conditions conducive to sustained population growth has also been widely questioned. Fisher (1964: 172) places considerable weight on the 'drop in the death rate resulting from the suppression of internecine warfare and the establishment of law and order by the Western power'. The violence associated with many of the annexations and the often protracted resistance that followed has been used to discredit this view. However, Reid (1987: 42–3) has argued more convincingly that there was a real decline in the incidence of events which dislocated production and gave rise to famine and epidemic diseases.

The real weakness in the whole debate is the paucity of evidence for a decline in the death rate beyond that implied from the increase in population. Indeed, as Owen (1987a: 50) has noted, in terms of hard information it is equally possible to argue that there was an increase in the birth rate. This has been suggested by a number of writers. White (1973), for example, has argued that in Java the cultural system, by placing a pressure on labour, caused the abandonment of traditional limits on fertility such as extended breast feeding and late marriages. Evidence for a fall in age of marriage has been advanced by Boomgard (1987: 280–1) for Java and Smith and Ng (1982: 248–52) for the Philippines.

Undoubtedly there is a close relationship between the incorporation of South East Asia into the world economy and the acceleration of the rate of population growth. The extent to which this can be attributed to changes in fertility or mortality remains very much an open question.

Spatial pattern of economic activity

The changes in population, however, are only one aspect of a much broader change in the distribution of economic activity. In the NEI Dutch annexation and colonial policy greatly enhanced the long-established imbalance between Java and the other islands. The legacy of Dutch policy in the NEI has been summarised by Fisher (1971: 296).

> Indonesia illustrates with unique intensity what is probably the greatest problem facing all the newly independent [South East Asian] states, namely the aggravation, and in this case morbid distortion by colonial rule of pre-existing regional differences within a territory.

Nawani (1971: 165) has also argued that the long period of Dutch rule broke down the interrelationships between the various islands and isolated them from each other and from the outside world producing 'economic stagnation and social ossification'.

The concentration of Dutch activity in Java sharply increased the

Table 4.9 *Value of exports, Java versus the Outer Islands, 1870–1940*

	Indexes		Percentage of NEI exports	
	Java	Outer Islands	Java	Outer Islands
1870	100	100	87	13
1900	172	400	70	30
1930[1]	555	3671	44	56
1940	n/a	n/a	34	66

[1] All petroleum accounted to Outer Islands
Source: Calculated from Furnivall, 1944: 337 and 36–7

contrast between the inner and outer islands. From the 1870s Dutch enterprise began to move into parts of west and east Sumatra, in search of accessible, fertile land to establish specialised estate and small-holder production. A major hindrance to development in the outer islands was the shortage of local labour. Expansion of production depended almost exclusively on migrants from Java.

In the last decades of Dutch rule the whole balance of the NEI was changed (table 4.9). So, far from Java being the richest island in the chain, the outer territories, whose population had risen to nearly 25 million, were in 1940 producing two-thirds of the agricultural exports and practically the whole of the petroleum, tin and other minerals, while the congested and increasingly impoverished metropolitan island, by then containing some 50 million people, 'was becoming an economic millstone round the country's neck' (Furnivall, 1944: 56).

In the Philippines, as was discussed in 3.1, the Spanish established control over an area which lacked significant political organisation above that of the village, urban centres or clearly defined core areas. For three hundred years Spanish activity was concentrated in Central Luzon with the remote upland and most of the southern islands remaining outside of effective control until the late nineteenth century. The resultant striking imbalance was intensified during the American period with the accelerated development of the Spanish core not being significantly offset by transport developments and agricultural colonisation schemes in North East Luzon, Mindanao and Bohol.

In the Malay Peninsula the rapid development of tin and rubber produced a strikingly uneven pattern of development between the east and west of the Peninsula (map 4.1). Prior to these developments the contrast was slight and indeed in certain periods the east coast had been of more significance (Khoo Kay Kim, 1977a).

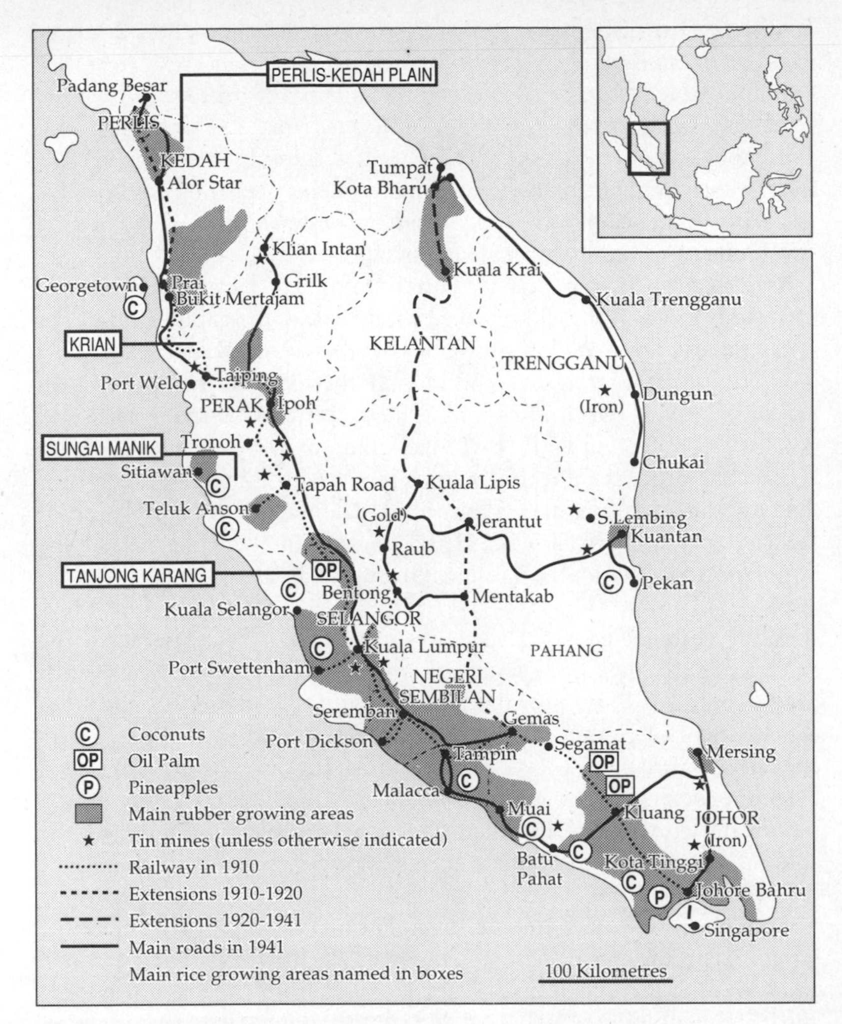

Map 4.1 Economic development in the Malay Peninsula, 1941
Source: Tate, 1979: 193

Initially the tin deposits nearest the west coast or on navigable rivers were exploited. As a result a series of small ports – Port Dickson, Port Weld and Port Swettenham (now Kelang) – grew up as trans-shipment points. To connect these centres and service the mining areas a rail network was constructed between 1885 and 1918. However, this infrastructure also opened up the interior to commercial agriculture. Thus rubber and, later, oil palm plantations were established in the west, reinforcing the disparity with the west initiated by the rapid development of tin mining. East of the Central Range large areas remained heavily forested, thinly peopled and the domain of shifting cultivators. Sedentary rice growing was long established in the river valleys and along the coast as were fishing communities. Despite various proposals for the development of the eastern side of the peninsula, by the late 1930s little had occurred beyond a scatter of estates and isolated mining developments, notably for iron and magnesium.

In Burma the focus of activity shifted from the Central Dry Belt to the Irrawaddy delta. Rangoon replaced Mandalay as the capital and developed rapidly as a major port-city. By the 1930s Rangoon and its industrial satellite at Syriam had become the focus of the national rail network built between 1877 and 1903 as well as of river, road and pipeline linkages. As Kirk (1990) has graphically described, Rangoon became the primary industrial, commercial and administrative centre of the new Burmese core area. Between 1891 and 1931 the population of the city increased from 176,583 to 406,415. In the same period the population of Mandalay declined from 170,071 to 134,950 (Fisher, 1964: 439).

While colonial economic activity was heavily concentrated in the delta lands, following the annexation of Upper Burma in 1885 British capital began to flow into the Dry Belt. Initially this encouraged the development of peanut and cotton production and related processing concerns, particularly at Mandalay and Myingyan (Kirk, 1990). From 1886 the beginning of oil production at Chauk and Yenangyaung significantly increased the importance of the old core area to the colonial economy but did little to redress the increasing dominance of the national economy by the delta lands.

The concentration of colonial economic activity in the delta and, to a lesser extent, the Dry Belt greatly increased the inequality between these areas and the peripheral upland areas. Apart from enclaves of forestry and mining, notably of zinc, lead and silver at Bawdwin and of tungsten at Mawchi, both in the Shan States, the hill areas remained thinly peopled and dominated by shifting cultivation. These areas, known successively as 'Backward Tracts' and 'Excluded Territories', were never fully incorporated into the colony.

The focus of the French colonial economy in Indo-China was in the long densely peopled Tonkin area. From 1902 Hanoi became the colonial capital and the town and its outport at Haiphong were substantially rebuilt and extended. While the importance of Tonkin increased through the French period, the colonisation of the Mekong Delta for rice cultivation, the growth of rubber plantations on the fringes of the delta and the development of Saigon-Cholon (Ho Chi Minh) produced a southern core that to a degree balanced that of the capital. Unlike Burma, Thailand and the Philippines, Hanoi did not become a primary centre (Kirk, 1990). In 1939 Hanoi had a population of 149,000 as against 110,000 for Saigon. Thus the regional contrasts within Vietnam were perhaps less stark than in many other parts of South East Asia. Elsewhere in Indo-China with the exception of the extension of commercial rice and maize into the Cambodian delta areas and the expansion of the old capital Phnom Penh, development was extremely limited and contrasted sharply with Vietnam.

In Thailand the distinction between the 'inner regions' most firmly under the control of Bangkok and the more loosely controlled 'outer regions' of the North, North East and South became increasingly sharp after 1855. Before the mid-nineteenth century, poor communications and self-sufficient village economies produced a relatively even pattern of development in Thailand (Dixon, 1981: 69). From then on the traditional economy began to break down in the more accessible areas. From the 1890s marked differences in the level of development of the parts of the kingdom were noted (Smyth, 1898; Grahame, 1924; van der Heide, 1906) and, after 1930, statistically measured (Zimmerman, 1931 and 1937; Andrews, nd, *c.*1936). However, while the central plain gradually emerged as the core of the export economy, the gradual incorporation of the North East into the international rice market, the development of teak and commercial rice production in the North, and tin and rubber development in the South resulted in a perhaps more diffused pattern of development than was the case elsewhere in the region.

Thus with the imposition of colonial rule the spatial pattern of economic activity in South East Asia underwent major change. New core areas emerged most notably in Luzon, the Irrawaddy delta and the western part of the Malay Peninsula, established cores were intensified and overall the disparities between cores and peripheral areas greatly enhanced. In this process a key element was the development of major urban centres.

Urban centres

Western penetration of South East Asia significantly modified the long-established pattern of urban centres. As McGee (1967: 52) has concluded, the first three hundred years of European contact resulted in the creation of an embryonic colonial urban network. Only in Java and the Philippines were urban centres geared towards administration and control over production; elsewhere they remained largely peripheral to the internal economies and provided bases for the establishment of European control over extra-regional trade. As was discussed in chapter 2, the expansion of trade after 1450 resulted in the growth of South East Asian cities. Where a permanent European presence was established, as at Malacca, this era of mercantile urbanisation set a distinctive and lasting impression on the urban form (Drakakis-Smith, 1990).

The present-day urban system in South East Asia is, however, very largely the product of nineteenth-century industrial colonialism. All the leading cities, with the exception of Bangkok[42] are essentially colonial creations (Fisher, 1964: 185). The major port-cities of Bangkok, Jakarta, Manila, Rangoon, Saigon-Cholon and Singapore were developed on the sites of existing settlements but, with the exception of Bangkok, these were small and largely peripheral to the pre-colonial economy. While Bangkok was already a major centre before western penetration in 1855 its subsequent growth was very largely a product of the same economic processes that operated within the formal colonial territories elsewhere in the region.

Following the imposition of colonial rule the growth of the port cities was rapid. Rangoon grew from a small trading port in 1852 to a town of 46,000 in 1855, 92,301 in 1872 and 406,415 in 1931 (Spate and Trueblood, 1942). As McGee (1967: 56) has stressed:

> The chief characteristic of these great cities, apart from their considerable size and growth, was their multiplicity of function. The most prominent function of these cities was economic; the colonial city was the 'nerve centre' of colonial exploitation. Concentrated here were the institutions through which capitalism extended its control over the colonial economy – banks, the agency houses, trading companies, the shipping companies and the insurance companies.

Thus these cities not only rapidly became by far the largest urban centres in their respective territories but also established a virtual monopoly of commercial administrative and manufacturing activities (Ginsberg, 1955: 457). They became a focus of transport routes, often to the detriment of integration of the national economy.

Such urban primacy was most striking in Burma, Thailand and the Philippines, where the port-capitals were by the 1930s at least five times the size of the next largest settlement.[43] In the Malay Peninsula and Indonesia political divisions produced a less-concentrated pattern; in Indonesia the size and diversity of the territory resulted in the development of a greater spread of urban functions. However, particularly after 1920 the growth of other centres, especially such major inland centres as Hanoi, Kuala Lumpur and Surakarta, tended to slow the rate of growth of primacy and in Indo-China actually reduced it.[44]

The dominance of the urban systems of these major centres has tended to obscure the significance of the proliferation of small towns. In the Malay States the number of urban centres (as defined by the census) increased from 92 in 1911 to 176 in 1931 (Caldwell, 1963; McGee, 1967: 55). Settlements established or expanded to serve as minor ports, rail junctions, mining, collection and administrative centres, performed a vital intermediate role between the Western-orientated port-cities and the internal economy. Thus the whole urban system became orientated as new or selected centres were expanded to cope with the changing structures of commodity exploitation (Drakakis-Smith, 1990). In the Malay Peninsula Lim (1978) concluded that an entirely new urban hierarchy resulted from the penetration of the tin areas by the railway network.

The major urban centres established during the colonial period formed the focus for the development of capitalist relations of production. It was through the major port-cities that the region became firmly locked into the international division of labour as a major supplier of raw materials. However, it was largely through the network of smaller urban centres that colonial rule was imposed, direct control over production established and capitalist relations of production diffused.

Class formation

By the end of the colonial period the basis of capitalist production had been laid in South East Asia. While the majority of the population remained rural the cores of urban-industrial structures had been established. By 1940 indigenous capitalist and working classes had emerged. While they remained small, poorly organised and, in consequence, politically weak, they were to play decisive roles in the immediate post-war period.

There is a tendency to underestimate the size of the 1930s South East Asian working class. Those employed in the European-controlled manufacturing, mining and plantation sectors were only part of it.[45] In addition, as Buttinger (1967: 194) has suggested, the 'tens of thousands' employed by local capitalists should be included:

> The Chinese owned many rice mills in Cochinchina and a considerable portion of all retail trade in Vietnam. The Chinese and Vietnamese also controlled all small river and road traffic. Printing presses, potteries, cabinet making, some sugar refining, soap factories, and other small establishments producing for the

> local market were almost exclusively in Vietnamese and Chinese hands. Hundreds of Chinese and Vietnamese owned stores, restaurants, and repair shops. All cities had Chinese and Vietnamese tailors, pharmacists, butchers, and many other independent suppliers of goods and services, and most of them employed some people. There is no reason why the many tens of thousands working in Chinese mills or Vietnamese shops should not be regarded as members of the working class.

It is also likely that official statistics underestimated employment in the European-controlled sector[46] and excluded large numbers of landless or near-landless labourers employed on a casual temporary or seasonal basis in the agricultural sector (Thompson, 1947: 171; Buttinger, 1967: 194).

The size of the non-Western capitalist class has similarly been generally underestimated because of the tendency to recognise only those activities that mirrored the European-controlled sector. Thus the extent to which South East Asian capital (including that of Chinese and Indian immigrants) was establishing itself in a wide range of service, trading, manufacturing and primary activities which supported the European sector and its associated workforce has been seriously under investigated.

In general, South East Asian capital was restricted to those areas unattractive to European activity. Capital accumulation took place primarily in mercantile and simple processing activities. However, Resnik (1973) has shown how in the Philippines, where very large estates had been established, the landed classes were able to use the proceeds of commercial agriculture to become urban merchants. In the early twentieth century some of this capital from trading and directly from family estates moved into agricultural processing and manufacturing, printing and construction. This direct link is certainly not the norm in South East Asia; more commonly trading capital moved into land ownership. This was particularly apparent in Thailand and Burma during the 1920s and 1930s.

Poverty

The incorporation of South East Asia into the world-economy not only created a more uneven pattern of population and economic activity but also set in motion the sharp differentiation of the region's communities. A direct consequence of this was the creation of mass poverty, attested to by a number of surveys conducted during the 1920s and 1930s.[47]

In Java the Dutch colonial system 'locked' the majority of the rural population in the highly subsistence rice sector. As Geertz (1963: 69) has shown, the Javanese were vital to the development of the estates and cash-crop production but they were never able to become properly part of it, 'it was just something they did, or more exactly were obliged to do, in their spare time'. The Javanese were largely restricted to the food-crop sector where, particularly after 1830, their numbers increased rapidly.[48]

Data on the Javanese agriculture sector are scarce until the 1920s. Rapid increases in population were accompanied by expansion of the cultivated area and conversion of shifting cultivation to increasingly intensive wet-rice production. The wet-rice sector was able to absorb almost all the vast increase in population created by the Dutch intrusion (Furnivall, 1944: 151).

By the end of the nineteenth century the intensification of wet-rice production was beginning to falter. The response was to plant dry-season food crops – maize, cassava and sweet potatoes – in the paddy fields. This practice spread rapidly between 1900 and 1930, maintaining per capita food production (but not necessarily consumption) at, or near, the late-nineteenth-century level (Geertz, 1963: 79). However, van de Koppel (1946: 369) concluded that per capita calorie intake dropped sharply from the late 1920s.

Geertz (1963) depicts the communal element in Javanese rural society being 'put into over-drive'. Communities were turned in on themselves and through a process of 'involution' a situation of mass shared poverty was created. While few would argue with the view that the majority of the Javanese population were greatly impoverished by the 1930s, the trends in living standards since 1800 have been the subject of considerable debate.

While some writers, such as Carey (1979: 10) have argued for a continuous fall in living standards since 1800, the weight of evidence points to a more complex pattern. Booth (1988), in an excellent review of the disparate evidence, suggests a sharp fall from 1815; a gradual amelioration from the 1840s which accelerated after 1870, perhaps by the 1880s regained the level that prevailed in the 1820s; a steep fall from *c.*1885 to *c.*1907; a brief recovery until 1915; and a steady fall from 1915 to 1930. By the end of the nineteenth century the Dutch colonial administration was becoming aware: 'that the forced changes introduced in the indigenous farming economy of Java by colonial policies meant steadily decreasing living standards for the people of the island ...' (Hardjono, 1977: 16). In 1899, the Minister of Colonial Affairs proposed that the standard of living of the Javanese farmer be improved by 'education, irrigation and emigration' (cited Tempelman, 1977: 12). The so-called 'Ethical Policy' adopted in 1901 centred on the revival of the state's role in economic development. This stemmed from economic and political change in Holland.

By the turn of the century the main Dutch political currents were in favour of a colonial policy which restricted the activities of the major corporations and raised the living standards of the Javanese. While it cannot be denied that a strong element of 'moral concern' over the colonies emerged in this period, the interests of Holland were recognised as being best served by raising living standards in the NEI and spearheading development in the outer islands by the state (Tate, 1979: 99). The establishment of health, welfare and educational facilities improved the environment for the Dutch businessmen, planters and speculators as well as providing them with a healthier, locally trained pool of labour. In addition increased real incomes in the NEI would expand the markets for the emerging Dutch manufacturing sector – particularly textiles (van Niel, 1960: 291).

The policies implemented between 1901 and 1941 promoted irrigation, education, rural credit, health measures and emigration from Java to the outer islands. During this period the irrigated area in Java increased by 24 per cent, a network of banks and pawn shops gave villagers access to cheaper and more secure credit than they could obtain from moneylenders,[49] and some 200,000 Javanese were resettled, principally in the Lampung Residency of Sumatra.

Irrigation works were built by the large-scale use of corvée labour, farmers being liable for up to a month each year. Once schemes were completed cultivators were obliged to sell much of their produce in order to meet the tax on irrigated land. In the nearby island of Lambak tax absorbed 20 per cent of the crop in 1914–20, 15 per cent in 1930–40 and 25 per cent in 1940–50 (Bray, 1986: 180). Where there were few alternatives to wet-rice cultivation this policy seriously depressed living standards.

Overall the Dutch colonial development policy provided only limited and localised relief for the impoverished population of Java.[50] Boeke (1966) concluded in 1927 that the welfare policies merely resulted in further population increase with no general improvement in the material well being of the population. While a case can be made for increased real income amongst some sectors of the community during the 1920s and 1930s (Polak, 1943) the weight of evidence (see Booth, 1988; Geertz, 1963; Fisher, 1964; Furnivall, 1944; O'Malley, 1979; Brown, 1986) still supports Kahin's (1952: 25) conclusion: 'even prior to the depression the general level of economic welfare [of the population] was declining, while it was being obliged to shoulder a heavier burden of taxes'. What ever the limited impact of the ethical policy on the living standards of the Javanese population transmigration played an important part in the opening of the outer islands. With the repealing of the 1880 'Coolie Ordinance' in 1930, state-sponsored migration became the major source of labour for the Sumatran plantation economy (Tempelman, 1977: 14–15). By 1940 there were 200,576 settlers in the various transmigration schemes (Hardjono, 1983: 56).

The effect of the exposure of an increasing proportion of the South East Asian population to fluctuations in the world-economy is most dramatically illustrated by the widespread distress observable during the 1930s following the commodity price collapse of 1929–31. While the inter-war recession, like early shallower and less protracted episodes, accelerated the process of class formation and impoverishment, it was not the root cause of the widely reported distress. It must be seen as exacerbating an already chronic condition. The causes of poverty were essentially structural rather than cyclical. Most studies of the recession infer a very direct relationship between the fall in commodity prices and poverty.[51] However, the impact of the inter-war recession was far from even.

Brown (1986) has argued that the effects of the inter-war recession may well have been exaggerated. The most commercialised areas and those most fully dependent on production for the world market and wage labouring suffered most (Dixon, 1981: 70). However, where a retreat to subsistence activities was possible, it may well be that per capita food consumption increased. In Java, however, the suffering of the urban population was very largely transferred to the rural areas, with fewer circulating migrants visiting the city for labouring jobs and many urban workers moving back to villages (Ingleson, 1988: 309). For those who retained jobs in the urban-industrial sector there is evidence of increased real wages during the 1930s but this has to be set against sharp deterioration in the conditions of employment (Ingleson, 1988: 309). In many areas O'Malley's (1979: 235) comment on 'how harshly the economic climate of the Great Depression affected a large part of the native populations' was all too accurate. The specialised tin and rubber areas of the Malay states were severely hit; in 1931 the number of unemployed Chinese miners was estimated at 450,000 (Khoo Kay Kim, 1977b: 82). In general the colonial administration made little provision for the unemployed. The Straits Settlement Legislative Council was typical in this respect: '[the] Malayan government [was] not satisfied that there [was] any need ... for the creation of special departments to deal with the unemployed'.[52] Further the Acting Colonial Secretary stressed: 'the government will not agree to any assistance in the shape of doles'.[53]

In consequence relief was left to a wide range of voluntary organisations,[54] while the colonial administration attempted to reduce public expenditure. In the Malay states the Retrenchment Committee, formed in 1930, cut government personnel by 50 per cent, while between 1929 and 1935 the Railway Department reduced staffing by 50 per cent and expenditure by 40 per cent (Khoo Kay Kim, 1977b: 88). In French Indo-China the colonial state not only reduced expenditure but increased tax levels to compensate for declining revenues (Scott, 1976: 119).[55] A similar increase in tax burden took place in Java (Booth, 1988) and possibly Burma (Scott, 1976: 119).

The 1930s were characterised by widespread unrest, particularly in rural areas. Most attention has focused on the major peasant rebellions: the Sakdal rising in central Luzon, the San Hsaya rebellion in lower Burma and the revolts in the northern Annam provinces of Nghe-An and Ha-Tinh. However, as indicators of rural distress, perhaps the scattered, small-scale outbreaks of violence that persisted throughout the period are of more significance (Brown, 1986: 995). These included raids on food stores, attacks on homes of landlords, merchants and moneylenders and the burning of police posts and government offices. Such incidents were probably most common in central Luzon and Cochinchina (Kerkvliet, 1977: 35–9; Scott, 1976: 120–7).

For the colonial administrations and their apologists the studies which linked South East Asian poverty to the inter-war recession, provided a convenient scapegoat for processes which since the late nineteenth century had made increasingly large numbers of the region's people vulnerable to changes in international demand for a limited range of primary products.

The 1930s: seeds of change

The 1930s were a formative period in the emergence of organised mass independence movements. While most of these movements dated from before the 1914–18 war they only crystallised into national organizations during the 1920s.[56] The movements were far from unified and covered a very broad political spectrum. This is perhaps most clearly shown in French Indo-China.

The Indo-Chinese independence movements emerged from the struggle against French annexation which continued with diminishing vigour until 1900 (Ramachandra, 1977: 244). In 1923 the Constitutional Party was formed and, in 1927, the Vietnamese Nationalist Party (VNQDA) which had many similarities with the Kuomintang. Both of these organisations were essentially nationalist movements which represented the interests of the emergent but increasingly frustrated Indo-Chinese capitalist class and as such lacked mass support. In 1929 three left-wing groups came together to form the ICP (Indo-Chinese Communist Party). Following the abortive 1930 rebellion both the nationalist and communist groups were heavily suppressed and their leaderships decimated. During 1931–2 over 10,000 were killed and 50,000 deported (Pirani, 1987: 28). Between 1933 and 1937, while a number of radical and nationalist groups remained active, the leadership in the struggle against the French passed to a 'united front' comprising the Left Opposition Group (the main Trotskyist element) and the remnant of the pro-Moscow ICP. This grouping was broken up during 1937 at the behest of Moscow (Sacks, 1960: 136–41). The ICP then formed a broad united front that included the nationalist groups. The Trotskyist groups, however, remained extremely active and retained considerable mass support. An increasingly bitter struggle was waged between these factions which continued into the immediate post-war period.

In general the colonial authorities tended to publicly dismiss the nationalist movements as a handful of hotheads not to be taken seriously.

> Undoubtedly this situation was aggravated by the astonishing ignorance on the part of individual colonial administrations about what was going on in each other's territories. For while in the mental isolation of particular colonies – all of them relatively small by Indian standards – the situation might appear to be well under control, seen as a whole it presented a much less stable appearance, and notwithstanding local differences of emphasis and pace, events were everywhere moving in the same direction. (Fisher, 1971: 251)

However, it is important not to underplay the degree to which the nationalist movements influenced colonial policy. From the early 1900s, in Burma, Indo-China and the NEI to varying degrees, the non-European population became involved in local and centralised administration. This of course, also reflected the emergence of various 'liberal' views in the metropolitan countries as well as administrative convenience and cost.

In the NEI, while the 'ethical policy' introduced after 1900 can very broadly be seen as preparation for a limited degree of self-government, the Dutch view was of very slow political evolution (Harrison, 1963: 242).[57] The involvement of non-Europeans in administration was extremely limited. In 1938 92.7 per cent of all government posts were held by Europeans (official Dutch figure cited by Golay *et al.* 1969: 134). Despite moves towards treating Java as a self-contained administrative unit whose interests might not always be the same as those of Holland, self-government was something for the very distant future. Indeed 'without undue scepticism the doubt may be expressed whether the Dutch government had the serious intention of ever granting Indonesia real self-government' (Hall, 1981: 796).

In French Indo-China from the early 1900s non-Europeans were recruited into the administrations and consultative chambers established. However, 'self-government was never the aim of French policy; assimilation rather than association was its keynote' (Hall, 1981: 798). Far fewer non-Europeans were recruited into the government service than was the case in the NEI and more especially Burma (Hall, 1981: 799).

While the role of the indigenous population in administration and decision-making was most restricted in the French territories, the façade of 'native administration' was an imposing one.[58] In theory only Cochinchina was a directly ruled colony; the other divisions were protectorates. Thus in Annam, Cambodia and Laos the kings, their courts and the mandarin structure remained intact. However, French control over this apparatus was absolute (Hall, 1981: 304).

In Burma non-European involvement in administration and decision-making was closely related to that of India.[59] The lower-level administration posts were principally filled by Burmese and, more significantly, Indians. After 1923 non-Europeans were recruited to the highest levels of the civil service and police. In that year Burma, excluding the Shan States, Karenni and the other 'tribal territories' became a 'governor's province' with a partly elected legislative council. While this body became responsible for forestry, education and public health, it had no control over defence, law and order, finance and revenue. In addition, rural and municipal committees, the majority of whose members were elected, were given a wide range of responsibilities. In sum, between 1923 and 1937 Burma achieved a very real degree of self-government in local affairs as well as at the centre (Hall, 1981: 781). By 1940, 70 per cent of senior civil service posts were held by Burmese, though British officers still dominated at the highest level and a large proportion of teachers, university staff, medical officers and post office employees were Indian (Golay *et al.* 1969: 259–60).

In 1937 Burma was separated from India and given almost complete control over internal affairs (Hall, 1981: 785). However, unlike India, no clear promise of full 'self-government' was made. Indeed the nationalists treated the separation, supported as it was by British business interests, with grave suspicion. In this they were almost certainly correct. The value of Burma to the British imperial economy made full independence unthinkable.[60]

By contrast, in the fragmented British territories in the Malay Peninsula and Borneo, where there was little in the way of an organised nationalist movement, the non-European population was largely excluded from the colonial administration.[61] It was the 1939–45 war and the Japanese occupation that was to give rise to a mass Malay nationalist movement (Akashi, 1980).[62]

The rapidity of post-war decolonisation is all the more striking in the light of limited and uneven progress towards self-government made by the late 1930s. The inter-war period was one of rapid change for the South East Asian states. Changed international circumstances brought deep-seated discontent to the surface and provided the conditions for the establishment of independence movements with mass support.

The 1914–18 war brought Britain's financial supremacy to an end and seriously weakened the French and Dutch imperial structures. During the inter-war period these powers were forced to fight a rearguard action against the attempts by the USA and Japan to penetrate their spheres of activity in South East Asia.

The increased dependence of the USA on imported raw materials was an area of weakness because of the lack of control over sources. This was particularly the case for tin and rubber, the UK and Holland effectively controlling world supply. Measures aimed at protecting the interests of the colonial powers, for example Britain's Imperial Preference Scheme adopted in 1932, further restricted the access of the USA to South East Asian raw materials and markets.

The role of the USA in the struggles that surrounded the establishment of these commodity restriction schemes illustrates both the importance of access to the raw materials for the American economy and provides an insight into the formation of American thinking about a more acceptable international economic order (Caldwell, 1978: 9–10). Thus the USA's role in post-war South East Asia stemmed directly from the identification of the region as vital to the prosperity of the USA.

Despite the restrictions imposed by the European colonial powers Japanese involvement in South East Asia increased substantially during the 1920s and 1930s.[63] During the early 1930s Japanese goods and shipping were rapidly replacing Dutch in the NEI. Despite attempts to restrict trade to Dutch vessels the share of Japanese shipping in the Java-Japan trade rose from 46.4 per cent in 1929 to 75.1 in 1933 (Shimizu, 1988). The increased involvement of Japan in the region was accompanied by an influx of personnel.

Japan also actively encouraged the region's independence movements, recognising them as a potentially powerful force for the ending of colonial rule and the opening of South East Asia for Japanese activity. As Fisher (1971: 25) has written: 'Their slogan of "Asia for the Asiatics" drew an instinctive response almost everywhere except among the overseas Chinese whose mother country was already the victim of Japanese aggression.' By 1940 there were 24,000 Japanese in South East Asia,[64] the largest settlements being in the NEI, Malaya and the Philippines (Yuen Choy Leng, 1978: 163).

Thailand, because of its non-colonial status and the comparatively low level of development of the kingdom's considerable market and raw material potential, was a prime target. Following the Anglo-Thai treaty revision of 1926[65] Britain's position was increasingly challenged by the USA and Japan. By the late 1930s tin and rubber exports which had previously been exported through the Straits Settlement were going directly to the USA and Japan.[66] In addition, American investors and MNCs such as the British American Tobacco Corporation, were beginning to take an interest in establishing production behind the Thai tariff barriers established following the 1926 treaty revision.[67]

Outside of Thailand the USA and Japan were far less successful because of the protectionist measures enforced by Britain and Holland (see figures 4.5, 4.6(a) and 4.6(b)). However, by the late 1930s the two new powers were engaged in increasingly intense competition which heralded the Pacific War conflict (Caldwell, 1978: 11).

4.5 Decolonisation

Between 1946 and 1958 the formal colonial structure in South East Asia was effectively dismantled. It is important to separate immediate causes from changes in the world-economy which made formal control of territory not merely unnecessary but actually inimical to the interests of international capital. The colonial powers, most notably Britain, had lost their key position in world trade, finance and production. This role was assumed by the USA, with whom the interests of international capital became increasingly identified. The dismantling of colonial structures was an important element in opening up increasingly large areas of the world to American investment, trade and resource exploitation. From 1943 American policy moved decisively away from earlier commitments to restore the South East Asian territories to the European colonial powers (Strega, 1983). Indeed the future dependence of the USA on raw materials imported from the emerging independent countries of South East Asia was publicly stated as early as 1946 (McMahon, 1981: 143). The political, economic and moral pressure that the USA put on the colonial powers has to be seen in this context.

In the immediate pre-war period, despite the growth of nationalism and changing world circumstances that the colonial powers found themselves in, a definite programme leading to independence had only been established in the Philippines. The 1935 'commonwealth' constitution heralded a ten-year programme of increased Filipino involvement in government which would bring independence in 1945. In the event the islands became self-governing in 1946 following a crash programme of American reconstruction. The Philippines, however, remained firmly tied to American political and economic interests (5.2). Indeed as McCoy (1981) has argued, the Philippines achieved 'independence' but did not undergo 'decolonisation'; in the 1960s relations with the USA still remained substantially similar to those that prevailed during the 1930s.

Outside of the Philippines the Japanese occupation, by fostering nationalism, engendering organised resistance movements and destroying infrastructure, was a significant factor in the rapid post-war decolonisation.[68] As the Allied powers re-entered South East Asia during 1945 they were faced with states which, whatever their final relationship with the Japanese,[69] had been free of Western colonial rule. The reimposition of the formal colonial structure and the reconstruction of the economies was likely to be a difficult and costly operation. The resources of the European powers were depleted and inadequate for even domestic reconstruction. Thus they were to become materially heavily dependent on the USA and susceptible to its influence.

The nationalist movements played a decisive role in determining the pace of decolonisation in the changed world circumstances of the 1940s. Indeed in the case of French Indo-China and the NEI it was the nationalist movements that put decolonisation on the agenda at all. It is important to see these movements and their success as being both a product and a part of the transition of the world-economy that was clearly taking place in the 1930s but which the 1939–45 war accelerated. This is not to underplay the importance of general strikes, demonstrations and armed resistance which characterised the 1940s and, in the case of Malaya, the 1950s as well.

The dilemma facing the colonial powers is well illustrated by British policy towards Burma and the Malay states. A critical factor in this was the differential degree of war-time damage that the territories had suffered. The Burmese economy was effectively destroyed by the fighting of 1942 and 1944–5. In contrast the Malay states had not experienced anything like the degree of conflict or destruction. The tin mines, plantations and infrastructure were still largely intact, having played a significant role in the Japanese war economy (Akashi, 1980). Thus the rapid re-establishment of pre-war production was possible with limited resources.

Smith (1988: 33 and 36) has noted:

> Paradoxically, the ability of the British to restore their position in Malaya more rapidly than in Burma may have been due to the higher value which the Japanese themselves had placed on their occupation of the Malay peninsula. Apart from the four northern Malay States, which were returned to Thailand in October 1943, the Japanese appear to have made up their minds from the outset to incorporate Malaya and Singapore into their own colonial empire. As late as February 1945, Tokyo reaffirmed the principle that they would not be given independence in any form . . . Burma, on the other hand, was more distant from Japan and less vital to its long-term strategic and economic interests. Its importance was mainly as a key link in the strategic arrangements of the Western Allies, which the Japanese needed to control for the duration of the war in order to cut off supplies from India to China and if necessary to threaten India itself. From the beginning the Japanese allowed the Burmese a role in the Ba-Ho Administration inaugurated on 7 April 1942; and by January the following year they were already talking about independence. Burma's formal independence – albeit subject to continuing military occupation – was proclaimed on 1 August 1943. The fact that many Burmese leaders 'changed sides' towards the end of the war did not alter the fact that during the three years of Japanese occupation they gained a type of political and administrative experience which had been denied to them even under the constitution of 1935. Whichever political group emerged on top after the British returned, it would be impossible for the former colonial regime to 'turn back the clock'.

In addition to the practical considerations attendant on the re-establishment of British control over Burma the country's resources were no longer of the same significance to the British or indeed the world-economies. This was particularly the case with oil. The rapid development of the Middle East and American reserves meant that oil was no longer in short supply. Thus while the tin and rubber of the Malay states were in demand in Britain and were potentially earners of dollars through export to the USA, this was not the case with Burmese oil, rice and teak. Thus: 'In terms of the harsh economic realities of the post-war world, Britain could afford to let events in Burma take their course . . . and could not afford to do much more' (Smith, 1988: 48).

Burma

The granting of nominal independence to Burma by the Japanese in 1943 undoubtedly raised the expectations of the already well-developed nationalist movements. At the end of the war the independence movement was focused on Aung San, the leader of the Burma National Army (BNA), and the broadly based Anti Fascist People's Freedom League

(AFPFL). However, on their return in May 1945 the British stressed their commitment to Burmese 'self-government' within the British Commonwealth. In effect this meant the re-establishment of the 1937 constitution but with a new agreement on which areas of responsibility would be left to the British Government. This proposal was opposed by Aung San and the AFPFL who wished for complete independence. The granting of this resulted from major resistance by the nationalists, including a general strike in 1947 which paralysed the country. Important as these actions were, the rapid British withdrawal does have to be seen in the context of the broader considerations previously outlined and the growing chaos within Burma.

When Burma became fully independent in 1948 the work of reconstruction had scarcely started. The country was still far from fully under central government control.[70] In addition the British administration of 43 per cent of the country with 16 per cent of the population as 'Excluded Territories' had perpetuated the split between the Burmese and the other minority indigenous peoples, most notably the Karen. Separatist ambitions amongst these groups had been fuelled by British promises made in exchange for guerilla activity against the Japanese (Fisher, 1971: 290). From 1947 these areas were prepared to fight rather than come under Burmese control (Hall, 1981: 880). In addition, a variety of groups including the Burmese Communist party were violently opposed to parts of the independence agreement. The quasi-federal structure that was erected to incorporate these areas did little to satisfy them. In 1948 a major Karen rebellion broke out and a state of virtual civil war has prevailed since.

Netherlands East Indies

On 17 August 1945, with Japanese encouragement, a declaration of Indonesian independence was issued. The re-occupation of the NEI by Dutch and British forces from September 1945 was accompanied by a clear statement that 'Indonesian independence' was not recognised and no negotiations could take place with the war-time leader Sukarno because of his 'collaboration' with the Japanese (Legge, 1972: 149–80). The Dutch announced a vague programme under which the NEI would be developed as 'a partner in a kingdom of the Netherlands' (Hall, 1981: 891–2). The NEI were regarded as indispensable to the economic recovery of Holland and thus only a very limited degree of 'independence' was envisaged (McMahon, 1981: 92–3). This was unacceptable to the nationalist leaders, and the Dutch position was rapidly undermined by the escalation of organised republican opposition.

While reoccupation established Dutch control in Java and the eastern islands, negotiations with the nationalists were opened (Lee, 1981). These resulted in the proposed formation of a 'United States of Indonesia'. This would leave the Republic of Indonesia in authority over Java, Madura and Sumatra, and Holland in control of the rest. The two governments would cooperate in 'the establishment of a sovereign democratic state on a federal basis' (Hall, 1981: 893). An agreement was reached which implemented the proposal in March 1947. However, this would have left the Dutch effectively in control of the main export-producing areas (Fisher, 1971: 300). Neither side appears to have been satisfied and relations rapidly deteriorated. In May the first so-called 'police action' was launched. Dutch troops occupied key areas of Java, Madura and Sumatra. At the behest of India and Australia the United Nations Security Council intervened, arranging a ceasefire and the renewal of negotiations between the Dutch and the republican leaders. These discussions also broke down, and in December 1948, in a second police action, Dutch control was established over the remainder of the republican territory and the government imprisoned. The Dutch were unable to extend effective control outside of the main urban areas and in the absence of direct support by Britain or the USA, it is unlikely that full control could have been reestablished.[71] However, the police actions turned international opinion against Holland. Not only was military assistance not forthcoming but also the USA was prepared to withhold economic aid until the Dutch 'showed signs of willingness to enter into serious negotiations with the Republic' (Legge, 1972: 234). While the nature and extent of pressure on the Dutch has been a matter of debate (van der Eng, 1988),[72] Holland clearly felt betrayed by the USA, particularly given the support received by the French in Indo-China (Gerbrandy, 1950).

Renewed negotiations resulted in full independence in December 1949. While agreement was reached over the safeguarding of Dutch interests in Indonesia the relations between the countries were soured by the independence struggle. The immediate causes of discord were the question of Indonesia's monetary 'debt' to Holland, part of which had been incurred during the Dutch attempts at reimposing colonial rule, and the jurisdiction over Western New Guinea (Irian Jaya) (Palmier, 1962: 71–2; Vandenbosch, 1976). Relations were further strained by Indonesian suspicions of Dutch involvement in the Bandung 'incident'[73] and the participation of Dutch troops in the resistance by some of the eastern islands, notably Amboyna, (Ambon), to incorporation into the Republic (Palmier, 1962: 73–84). These events paved the way for the events of 1957 when Dutch concerns were nationalised, nationals expelled and Indonesia partially withdrew from the world-economy (see section 5.2)

The Malay States and the Straits settlements

The reestablishment of British rule in the Malay peninsula was a comparatively peaceful affair. However, the British seriously underestimated the extent to which nationalism and political organisation had developed in the territories during the Japanese occupation. During the war a number of plans had been drawn up for the reorganisation of the loose and complex administrative structure of the peninsula.[74] In 1946 the Union of Malaya was established. This comprised the Federated and Unfederated States, Penang and Malacca. The Malay Sultans retained their position but sovereignty was transferred to the British Crown (Andaya and Andaya, 1982: 255). It was originally envisaged that North Borneo and Sarawak would join the Union but in the event they became separate crown colonies.[75] Similarly, Singapore was deemed too important to British interests in Asia to be incorporated into the Union (Fisher, 1968: 118–19).

The inauguration of the Union was followed by such widespread, organised and effective opposition, particularly by the Malays, that the plan was never brought fully into effect (Andaya and Andaya, 1982: 259). In 1948 the Union was replaced by the Federation of Malaya. This was essentially a result of negotiations between the British, UMNO (United Malays National Organisation) and the Sultans. While a strong central government was established the sovereignty and individuality of the Malay States was maintained. The Federation represented a victory for the Malays and caused great dissent amongst the other ethnic groups, particularly the Chinese.

The wave of strikes and outbreaks of violence during 1946 and 1947 reflected the strength of nationalist feelings.[76] Behind these activities the main organising force was the MCP (Malay Communist Party) which dominated the GLU (General Labour Unions), which directly controlled 80–90 per cent of the unions (Stenson, 1970: 124). During 1948 the MCP moved to a position of armed struggle and in June a State of Emergency was declared (Andaya and Andaya, 1982: 258).

The 'Malay Emergency'[77] has been summarised as:

> a period of tension, instability and totalitarianism. The first quarter of this 12-year period saw intense communist guerilla activity, including the dramatic ambush and killing of the British High Commissioner, Sir Henry Gurney, on 6 October 1951; the second quarter, a lull in the number of communist guerilla activities or 'incidents' (clashes or contact between security forces and commun-

> ist guerillas) and the establishment of the totalitarian regime of General Sir Gerald Templar, Gurney's successor; the third quarter saw the British government transferring political power and independence to the country's non-radical nationalists; and the final quarter, the consolidation of the position of the country's non-radical nationalist élite as well as the economy and military strength of the British. (Cheah Boon Kheng, 1977: 111)

The escalating cost of British involvement,[78] disruption of production and investment and international opinion all played a part in the decision to find a method of withdrawal that would protect British interests.

Independence was granted in 1957 to a far from unified country. The separate units of the pre-war period were not fully integrated and the ethnic mix was potentially an explosive one. During the late 1950s and early 1960s a fragile unity was developed by the leadership (Fisher, 1968: 118).

Singapore, which had been excluded from the Federation in 1948, obtained internal self-government in 1959. Britain retained control of defence and foreign policy. This reflected the key position that the Malay territories had come to occupy in post-war British policy.[79]

Singapore, like Malaya, had been far from quiescent during the 1940s and 1950s. There was widespread opposition to the continuation of British rule. The level of strikes and unrest resulted in an exodus of British capital and activity (Fisher, 1968: 120 and section 5.2 this volume). There is little doubt that the latter hastened the relinquishing of control over internal affairs. These events reinforced the view that Singapore should be kept separate from the Federation.

Following the successful campaign of Lee Kuan Yew's PAP (People's Action Party) in Singapore's first elections, moves were made to stabilise the state through very repressive measures and to gain access to the Federation. Essentially the view of Lee Kuan Yew appears to have been that Singapore needed the market and resources of the Federation and would become the key industrial, entrepôt and financial centre within such a wider structure (see section 5.2). To a degree these aims were to be fulfilled in the shortlived – 1963–5 – Federation of Malaysia.

In many way the Federation of Malaysia can be seen as a piece of political arithmetic, with the inclusion of the predominantly Malay Sarawak and Sabah 'balancing' the Chinese population of Singapore.[80] In the event the ambitions of the PAP and the risk of racial conflict brought about the exclusion of Singapore from the Federation in 1965. Fisher (1968: 63) summarised the situation:

> With the benefit of hindsight it is easy to see the fallacies of the political arithmetic which underlay the separation of Singapore from peninsular Malaya in 1946, its continuation in 1948 and 1957, and the belated attempt to unite the

whole of the former British Malaysian sphere in 1963. Thus, in the first place, the very process of juggling with numbers, attempting to balance one community against another, with or without the aid of constitutional devices to achieve an equipoise, serves of itself to keep alive the sense of difference and of rivalry between the communities concerned. And secondly, numbers are of little relevance except when related to the facts of geographical distribution and economic function.

In the modern world, in which cities are the pace makers, and the bigger the city the faster the pace, the gap between the metropolis and the countryside is tending all the time to widen. And since, in contrast to the predominantly urban Chinese, the Malays remain even more markedly rural, it is not surprising that the inclusion in the Malaysian Federation of Singapore, geographically the summit of the urban hierarchy of the whole region, could not be politically counterbalanced by the somewhat smaller total of overwhelmingly rural Malaysians in Sabah and Sarawak, even though the latter were allocated a disproportionately large share of votes in the Federal Parliament. For the Borneo indigenes did not form a single group linguistically, nor did they automatically align themselves with the Malay vis à vis the Chinese.

Thus, notwithstanding the fact ... that the Chinese remained in a minority in Malaysia as they had been in Malaya, the situation was nevertheless fundamentally changed in that within Malaysia they were the largest single community, with 42 per cent, leaving the Malays in second place with 39 per cent. Moreover, even if Malays and Borneo indigenes were grouped together, they were still in a minority with 46 per cent, but the Chinese and Indians together comprised a majority of 51 per cent.

Thailand

The British control over Thailand which was receding during the inter-war period came to an effective end after World War II. Anglo–Thai relations were adversely affected by the rapid capitulation to Japan which opened the 'back door' to the Malay States and Singapore, and the subsequent, 'declaration of war on the allies'. In addition Britain had experienced considerable commercial loss in Thailand as a result of the war. On both these counts the British Government pressed for heavy compensation. The USA never recognised the Thai declaration of war and pressurised Britain into moderating the demands while advancing American interests in Thailand.[81] Thus, during the 1940s, American influence replaced that of Britain, a foretaste of what was to develop in South East Asia as a whole.

Indo-China

By the late 1930s the areas of French Indo-China that were to form Vietnam were among the most strongly nationalist areas in South East

Asia (Fisher, 1964: 553). As elsewhere in the region, the Japanese occupation greatly intensified these feelings. Unlike the other occupied areas the colonial administration was left largely intact (Kiyoka Kurusa Nitz, 1984; McCoy, 1981). The war-time French administration, in order to offset the reality of collaboration with the Japanese, fostered a form of Indochinese nationalism and allowed more Vietnamese to enter the middle and upper echelons of the administration (Marr, 1981: 193). However, in 1945, following the withdrawal of the Japanese regional headquarters from Singapore to Saigon, the French administration was removed. The end of the 'colonial status' of Indo-China was announced by radio broadcast (Smith, 1978). Declarations of independence were issued and three states established, Vietnam (occupying Tonkin, Annam and Cochinchina), Cambodia and Laos, to be ruled respectively by Bao Dai, the Emperor of Annam, and the kings of Cambodia and Luang Prabang.

In 1941 the long-exiled[82] Ho Chi Minh returned and coordinated the establishment of the broadly based Viet Nam Doc Lap Dong Minh (Vietnam Independence League), later simply referred to as the Vietminh. This was to be the main focus of opposition to the Japanese and the French 'collaborative' regime. As such it received American support and was supplied with munitions and military training.[83]

The Vietminh refused to accept the jurisdiction of Bao Dai. Following the Japanese capitulation they occupied Hanoi and Ho Chi Minh issued his own declaration of independence. Control was consolidated in the north but remained loose in the southern provinces and far from firmly established in Saigon.[84]

The Vietminh were a broadly based nationalist movement that had much in common with those elsewhere in South East Asia. Essentially it was a 'popular front' which protected the prosperity of the Vietnamese bourgeois (but not that of the French) and had a programme of 'national liberation' and agrarian reform. The domination of the leadership of such a movement by the Communist Party of Indo-China was, as we have seen, paralleled elsewhere in the region, and was of course fully compatible with the orthodox 'Stalinist' position on the two-stage revolution.[85]

The British occupation of the South during September 1945 paved the way for the return of the French. Following an 'orgy of violence' by French paratroops and Foreign Legionnaires on 22 September a general strike paralysed Saigon (Karnov, 1983: 149). Despite further negotiations and French offers of 'self-government' within the French Union,[86] the scene was set for a major conflict between the French and the nationalists.

The French made clear their intention to reoccupy through military force. This attempt was to take place with no apparent opposition from the USA. In contrast to the pressure exerted on the British and Dutch to decolonise, the French 'recolonisation' was tacitly, and later actively, supported. In the initial stages the USA considered that the retention of all or part of Indo-China would stabilise the French State (Gettleman *et al.*, 1985: 42–7; Kahin, 1986: 3–6). However, the rapid progress of the Chinese revolution, and the emerging 'cold war' brought Indo-China to the fore in American strategic thinking. The French were not defeating the Vietminh, which were increasingly perceived as 'communists' rather than 'nationalists'. Under these circumstances the USA was prepared to accept the position of the French right-wing that the Indo-China war was part of the global struggle against communism (Kalb and Abel, 1971: 59). This move was reinforced by the recriminations that followed the 'loss of China', the rise of McCarthyism and, in 1950, the outbreak of the Korean War.

From 1946 American military aid was shipped to France, thus indirectly supplying French forces in Indo-China. During 1950 a programme of direct military aid was established; this grew from US$500 million in 1951 to US$875 million in 1952 and US$1,33 million[87] in 1954 (Kalb and Abel, 1971: 67 and 75). Despite the massive volume of aid during 1954 the French forces were outmanoeuvred and fought to a standstill at Dienbienphu. The USA declined to salvage the French position by direct military action.[88]

The 1954 Geneva Conference provided for a military truce but produced no political solution. Vietnam was divided along the 17 degrees north line of latitude into the Democratic Republic in the north and the Republic of Vietnam in the south. Provision was made for national elections in 1956 which, it was clear to all, the Vietminh would win. Other than the cease-fire no agreement was formally signed; there was merely an oral endorsement which did *not* include the USA and the southern regime (Karnov, 1983: 204; Turley, 1986: 6). With the benefit of hindsight the Geneva Conference merely gave rise to a brief period of respite between two major wars.

The conference also confirmed the independence of Cambodia, conceded by the French in 1953. In Laos an unstable royalist government was recognised and admitted to the United Nations in 1955 as an independent state. Laos was a scarcely viable economic unit and this government never effectively controlled the country. An intermittent struggle continued with the Pathet Lao forces which controlled the two northernmost provinces of the country.

After 1954 French forces were withdrawn from the south and

replaced by increasing amounts of US economic and military aid. The south declined to allow elections in 1956 and from 1957 open conflict escalated rapidly.

The increasing involvement of the USA was part of that country's broader global and Asian strategy. As Smith (1983: 190) has emphasised there were three specific aspects to Vietnam's importance to the USA:

> First, as an American ally South Vietnam has received nearly US$1.4 billion in various forms of aid since 1955 and had become a test case for the concept of 'defense support', making it essential to prove that aid on such a scale was not in vain. This was not mere symbolism, for South Vietnam was by now the weak link in a chain of American alliances with anti-Communist states across Asia. Both sides knew that, given the strategy the Americans had chosen to adopt, the 'loss' of Vietnam would have serious practical consequences. Second, the fact that South Vietnam and Laos bordered directly onto the social camp gave them a special status in the American theory of 'containment'. Whatever the legal ambiguities of military partition, the United States was by now committed to respond to any unilateral attempt to reverse the status quo at the edge of the 'free world' – whether in Indochina, West Berlin or South Korea. Whilst the intrinsic worth of South Vietnam might be much less than that of the other two areas, the parallel had a decisive effect on the way its security was viewed by the Pentagon and the State Department. Third, Vietnam belonged to the Afro-Asian world and was a logical target for revolutionary armed struggle of a kind not easily countered by conventional military means. The success of Castro in Cuba, coinciding with the apparent change in the Communist line on 'national liberation' wars, meant that the Americans must find ways of meeting their self-imposed obligations to 'contain' Communist expansion.

Acceptance of these views should not blind us to the importance of South East Asian markets and resources to the world capitalist economy in general and that of the USA in particular.

An account of the last and most bloody of the nationalist conflicts in South East Asia is beyond the scope of the present book.[89] In addition to the destruction of all the former Indo-Chinese states the war has had a profound and lasting impact on South East Asia and its position in the world-economy.

The decolonisation of South East Asia resulted in a substantial proportion of the region withdrawing or being excluded from the world capital economy. In those countries that remained pro-Western, decolonisation left far from stable situations. During the 1950s and 1960s major revolutionary movements developed throughout South East Asia. With some justification could Osborne (1970) entitle his book *Region of revolt*. The retention of these territories within the capitalist world necessitated widespread Western, principally American, military involvement and the creation of a series of right-wing 'garrison states'.

5

Development strategies and the international economy

5.1 Overview of South East Asian development strategies

This chapter examines the development strategies followed by the capitalist states of South East Asia since the 1950s in the context of their changing interaction with the world-economy. Major changes in policy show considerable similarity in form and timing throughout the region. This reflects the generally 'open' nature of the capitalist South East Asian economies, their high degree of interaction with the world-economy and their receptiveness to changes in the prevailing development 'orthodoxy', in particular through the advocacy of the IBRD and IMF. However, this changing orthodoxy is itself a reflection of change in the world-economy and indeed trails behind such changes.

Since the 1950s economic development strategies in all the pro-capitalist South East Asian states have emphasised rapid urban-industrial development. In addition as Hardjono (1983: 33) has noted: 'insecure governments tend to concentrate their resources upon urban improvements in the interests of maintaining their own power and winning international recognition, of their country's status as a self governing nation'. Rural development has played an increasingly secondary role; although, in all the states except Singapore the rural areas have continued to receive a large proportion of development funds, it has generally been substantially less than their shares of the population and GDP.

Rural insurgency, particularly in Thailand and the Philippines and the strength of rural political lobbies, notably in Malaysia, have a significant part in maintaining expenditure on the rural sector. Where powerful rural elites exist they have been able to influence and generally benefit from development expenditure, thus reinforcing their position. In addition, the need to reduce the expenditure on food exports focused attention on the raising of agricultural productivity, most importantly in Java, Peninsular Malaysia and Central Luzon.

The rural sector has generally been treated as a supplier of cheap food and raw materials. This has been reflected in both policy and infrastructural developments. Capital accumulated in agricultural and related processing and trading has been transferred directly into manufacturing investment. This has been particularly evident in the Philippines and Thailand. Thus the rural sector has tended to subsidise urban-industrial development.

In Thailand the Rice Premium, which from 1955 kept the domestic price of rice up to 35 per cent below the international level, helped to redistribute income in favour of the urban dweller (Sithiporn Kridakara, 1970). The government view appears to have been that depressing the domestic price of rice was desirable because it held down the cost of living in urban areas, reduced pressure for wage increases, encouraged exports and enabled domestic industry to compete with imports (Myint, 1972: 31). Ingram (1971: 253) has suggested that this turning of the terms of trade against the rural sector played an important role in promoting economic development by transferring resources to the 'modern sector'. Policies of either allowing or encouraging, the movement of the terms of trade against the rural sector in order to support urban-industrial development are inherently contradictory. Rural poverty and low agricultural productivity are reinforced, thus restricting purchasing power and limiting the domestic market for manufactured goods.

As was discussed in chapter 4, all the post-independence South East Asian states inherited in varying degrees a legacy of uneven development, limited national integration and 'plural' societies. The political instability of this heritage has been all too apparent in the region. A high level of expenditure has been necessary merely to maintain sufficient stability for domestic and more especially foreign capital to operate. Malaysia has been by far the most extreme example of this (see section 5.2). The high level of defence and repressive 'internal security' expenditure discussed in chapter 1 has similar origins. During the 1960s and early 1970s much American funding combined 'rural development and communist suppression' in countries with rural 'security problems'. This was most clearly the case in Thailand and the Philippines (see section 5.2). In addition expenditure on rural development has reflected the extent to which rural interests have been politically represented.

Primary production

Immediately after the Second World War the general shortage of raw materials attracted domestic and, more significantly, a degree of international capital to reconstruct and develop primary production. This process was given added stimulus by the commodity boom associated with the Korean war. However, political instability, coupled with the attitudes of some post-colonial governments, increasingly restricted foreign investment to Malaysia, the Philippines and Thailand. The nationalisation in 1957 of Dutch-owned concerns in Indonesia left only the oil sector with any notable foreign involvement. In Malaysia some plantations were sold to local interests and a few were sub-divided into small estates (Aziz, 1962). However, by 1972, 52 per cent of the plantation sector was still under direct foreign control (Ooi Jin Bee, 1976: 214). While there has been internal pressure for 'indigenisation' (Courtenay, 1979), globally MNCs have been reducing their interests in the production of established plantation crops such as rubber. Direct control of agricultural production by MNCs has given way to indirect control through domination of related processing, transport, marketing and supply of inputs (Bradley, 1986: 98). In addition foreign capital has been attracted into a range of new primary and primary processing activities, notably bananas, pineapples, oil palm and fishery products. The continued importance of South East Asian primary production to the world-economy discussed in chapter 1 is illustrated by the rapid influx of foreign investment into Indonesian primary production (timber, oil, minerals and agriculture) after the installation of a pro-Western government in 1967 (see section 5.2). Similarly, following the oil price rise of 1973–4, multinational corporations rapidly developed oil and natural gas production in Indonesia and Malaysia.

Industrialisation

While all the newly independent pro-capitalist South East Asian states acknowledged the importance of primary production they all became committed to industrial development. There were a number of reasons for this. Firstly, it reflected the prevailing orthodoxy of development. Secondly, industrialisation was closely associated with development in the minds of independence leaders, just as primary production and the import of manufactured goods was associated with backwardness and colonial status. In many instances demands for industrial development had been associated with the independence movements. Thirdly, the economic structures of colonialism had confined the rising indigenous capitalist class to trading, limited non-agricultural primary production, and processing.

Overall a policy of industrial development appealed to nationalist sentiments and was seen as 'radical' in that it challenged the 'law of comparative advantage' which was used to justify colonial and former colonial countries' continued role as primary producers. Further stimulus was given to this by the sharp fall in commodity prices in the late 1950s and a general deterioration in the primary producers' terms of trade.

The industrial development strategies that were implemented from the 1950s were, with limited exceptions, ISI (Import Substituting Industrialisation) aimed at the reduction of dependence on imported goods. Inevitably these 'infant industries' needed to be protected from foreign competition. In all the South East economies a complex structure of tariffs, import quotas and bans, incentives and subsidies has been established. The lack of indigenous capital dictated that in most of the South East Asian countries the state, together with foreign enterprise and capital, would play a major role in industrial development. Thus in all the countries, to varying degrees, alliances of state, foreign and domestic capital were forged.

In all South East Asian countries the initial phase of ISI resulted in comparatively high rates of growth. This reflected both the very low base from which industrialisation was starting and an unsatisfied domestic demand for, in particular, consumer goods. However, by the early 1960s the ISI sectors were beginning to falter (except in Indonesia where the development was still embryonic), as the limitations and inherent contradictions of the strategy began to surface.

By the mid-1960s there had been little development apart from assembling of consumer goods for which the local markets were approaching saturation. There were few signs of a transition to capital goods production. In general manufactured imports were merely replaced by raw materials, capital goods and components. In addition, ISI became increasingly criticised both at the national and international level because protection fostered inefficient industries which would never become internationally competitive. Pressure for accelerated growth through the development of export markets was emerging in government, international agency and indigenous capital circles. From the mid-1960s, first in Singapore, development strategy began to move towards EOI (Export Oriented Industrialisation). By the early 1970s EOI had become the 'new orthodoxy' urged on Third World countries by the World Bank (Meier, 1976).

The change from ISI to EOI was not merely the product of 'rethinking' development strategy but stemmed directly from profound changes in the world-economy that were to give rise to the NIDL (New International Division of Labour). In essence a number of Third World countries were in a position to seize the opportunities offered by these changes. Their apparent success in accelerating economic growth on the basis of the export of manufactured goods, resulted in the approach being widely advocated.

However, as Robison *et al.* (1985: 7), have noted, the countries of 'South East Asia were not automatically subsumed into the NIDL'. ISI continued to play a major role. In part the continuation of ISI can be explained by inertia and vested interest, particularly where it was related to a complex and entrenched bureaucratic structure. Similarly some countries' involvement was limited by the assessment by foreign investors of their stability and profitability. In addition the changes in the world-economy that fostered EOI paradoxically also created conditions that made it possible for some countries to continue with ISI.

In the early 1970s international capital, including 'aid' and 'soft loans' channelled through agencies such as IBRD and UNDP was available for ISI and associated infrastructure. American involvement in South East Asia was important in maintaining capital flows into countries and sectors to which foreign investment was not attracted. This was particularly the case in Thailand and the Philippines. More significantly, particularly after 1973, the availability of surplus capital, mainly recycled petro-dollars, made it comparatively easy for countries to borrow heavily for both EOI and ISI. In the case of Indonesia and Malaysia import earnings from oil and gas sales were also available to finance ISI.

Thus while EOI spread gradually through South East Asia from the late 1960s it only became the dominant policy in Singapore. Elsewhere, to varying degrees, ISI was maintained. The continuation of these two strategies has left South East Asia with a complex and, often, conflicting pattern of industrial protection and export promotion.

In general protection has tended to increase since the 1960s in all the South East Asian countries except Singapore. Given the enormous range of tariffs and goods involved generalisation over levels of protection are far from easy (see De Rosa 1986; Findley and Garnaut, 1986; Ariff and Hill, 1985). Table 5.1 gives a general indication of the variations in protection level found in the region. For individual key items very high barriers were established; for example, the effective level of protection for consumer durables in Thailand was 495.6 per cent in 1978 (Akrasansee 1980). Despite moves towards 'trade liberalisation' in the 1980s,

Table 5.1 *Tariff protection in South East Asia, 1978*

	Mean tariff level	Mean tariff [1] weighted by value of imports	Effective Production [2] for manufacturing sector	% of items with tariffs above 30%
Indonesia	33.0	20.2	n/a	40.0
Malaysia	15.3	6.6	55.0	14.0
Philippines	44.2	23.0	70.0	40.0
Thailand	29.4	30.4	70.0	3.0
Singapore	5.6	3.7	n/a	24.0
ASEAN	25.5	20.9	n/a	n/a

[1] This under-estimates protection levels when a high tariff curtails imports.
[2] This takes account of tariffs on imports and the valued added within the country.
Source: Compiled from Kraus and Lütkenhorst (1986): 26–44

largely at the behest of the IMF and World Bank, there has been little effective reduction in levels of protection.[1]

In addition to tariffs a wide range of quota, import licensing and customs controls have been established. Particularly cumbersome import procedures can act as effective 'unofficial' barriers. This has been widely reported by companies operating in Indonesia.

There has been little development of EOI which makes use of domestic raw materials. Policies aimed at encouraging foreign investment into these sectors have not been notably successful though, it must be said, they have with few exceptions not been vigorously pursued. In general, the structure of tariffs and freight rates together with established developed-world MNC control over distribution and processing have limited these developments to simple processing of established primary commodities, for example rubber, tin and timber and newer ones such as oil palm, fruit, vegetables, meat and fishery products. Exceptions included the Indonesian wood-processing sector and the canning and freezing of sea-foods in Thailand. The main areas of EOI development have been domestic- and foreign-funded labour-intensive manufacturing which remains heavily reliant on imported inputs.

With the adoption of export promotion strategies by all the South East Asian states since the 1960s the various promotional measures have come to look very similar (table 5.2). There is, of course, some variation in detail and intensity. Whether these differences are of any significance in the attraction of MNCs and foreign investment or in developing an indigenously controlled export sector is a matter of debate. Certainly there is increasing rivalry between countries based on presenting an attractive 'package' to foreign investors. There is, however, little doubt

Table 5.2 *Export incentives*

	Indonesia	Malaysia	Philippines	Singapore	Thailand	Hong Kong	South Korea
1. FINANCIAL INCENTIVES							
Loans/interest reductions	x	x	x	x	x		x
Guarantees	x	x	x	x		x	x
2. FISCAL INCENTIVES							
(a) Tax exemptions and relief	x	x	x	x	x	x	x
(b) Depreciation allowances	x	x	x	x		x	x
(c) Exemption/remittance of customs duties and taxes on imports	x	x	x	[1]	x	[1]	x
3. FACTOR INCENTIVES							
(a) Training				x			
(b) Research and development				x			
(c) Sites, buildings facilities	x	x	x	x	x	x	x

[1] Irrelevant because of almost free trade regime.

Source: Kraus and Lütkenhorst, 1986: 45

that the measures are extremely costly when considered in terms of tax foregone, subsidies and direct expenditure.

In 1983 ESCAP (Economic and Social Commission for Asia and the Pacific) produced a study of the incentive programmes of twenty-eight countries.[2] This report commented on the high cost of promotional measures and concluded that 'the evidence of these studies does seem to point towards the ineffectiveness of these schemes in inducing new investment. The most that can be said is that their impact is slight or unknown' (ESCAP, 1983: 65). Other studies have concluded that 'incentives are only able to *reinforce* primary investment motives (e.g., low labour costs, high capacity of the domestic market), but not take their place. In many cases the answer is that generous incentives are regarded as an *indicator* for a host country's positive attitude towards foreign investment' (Kraus and Lütkenhorst, 1986: 50).

Similar criticisms have been made of the widespread establishment of individual estates and EPZs (Export Processing Zones) (ESCAP, 1983). Whereas EPZs may act as 'shop windows' for promotional activity in general, their direct benefit to the local economy is limited. Plants locating in EPZs develop few links with the rest of the economy. Thrift (1986a: 55) concluded that in most cases the zones are of no more than regional importance; their only real local linkages were through service sector activities such as banking, transport, power supply and maintenance. In terms of the generation of export earning it may well be that EPZs run at a loss. Morello (1983), in a study of three zones in the Philippines, found that exports in 1981 totalled US$ 236.2 million and imports US$ 173.7 million. The 'gain' of US$ 62 million does not, however, take into account interest, royalties, licence fees and profit repatriation. Similarly the zones have generated comparatively little employment – perhaps 73,000 jobs in Malaysia[3] and 105,000 in Singapore which were by far the largest developments (Thrift, 1986a: 53). The majority of this work is temporary and poorly paid; there is little evidence to support the view that EPZs lead to the transfers of skills or technology into the host country. ESCAP has been highly critical of EPZs, describing them as 'costly public outlay in support of private enterprise' (ESCAP, 1983: 63).

Producer services

The development of 'internationalised' manufacturing sectors has been closely associated with the expansion of 'producer services'. However, while attention has focused on the growth of the service sector (see chapter 1) in general this aspect of the internationalisation of production and finance has received comparatively little attention. Yet it has become an increasingly significant element in South East Asian development, particularly in Singapore, which is increasingly challenging Hongkong as the leading financial centre (see section 5.2). The establishment of a full range of producer services in major centres is both a consequence and a facilitator of MNC activity and foreign investment.

Producer services in South East Asia have been dominated by EEC-based concerns rather than the USA and Japan (Thrift, 1986b: 143). However, all the region's economies have been attempting to develop their own sectors. To this end an element of protection has appeared. For example, in Malaysia the passing of the Insurance (Amendment) Act of 1982 restricted the activities of foreign concerns. In consequence Lloyds, previously very active in the country, has withdrawn entirely (Murrey, 1986: 154). However, it seems unlikely that producer services will become a significant source of growth outside of Singapore.

5.2 South East Asian states and the international division of labour

Singapore

The process of internationalisation which all the pro-capitalist South East Asian states have been subject to is shown in its most extreme and consequently clearest form in Singapore. Additionally this 'high-point' of the internationalisation of a Third World economy reveals most fully the contradictions inherent in this development. In these respects a detailed account of Singapore provides a 'yard-stick' against which the other states may be reviewed.

Of all the South East Asian economies Singapore has become by far the most fully integrated into the NIDL. Miraz (1986: 2) has described the city-state as 'one of the few truly internationalised economies in the world'. It is to this openness and integration that many have pointed to in explaining Singapore's remarkable rate of development since the mid-1960s. However, the transformation of the Singapore economy has not been a product merely of openness to the world economy; development has been planned and directed by one of the most powerful and wide-ranging state apparatuses in the Third World (Miraz, 1986: 49). The Singapore state was in a position to discern and seize upon the opportunities offered by the changing structure of the world-economy.

By the 1950s, although Singapore remained primarily an entrepôt, a broad industrial sector had developed based on processing, associated light engineering, assembly of vehicles, marine engineering and a range of consumer goods (table 5.3). A beginning had been made in the

Table 5.3 *Singapore, employment in manufacturing and processing industries, 1955*

Industry	Number employed	Percentage share
Shipbuilding, repairing and marine engineering	10,813	43.8
Engineering works and manufacture of non-electrical machinery	4,767	19.3
Motor-vehicle repairs	4,540	18.4
Printing and publishing	2,291	9.3
Rubber milling	2,281	9.2
Total	24,692	100

Source: Colony of Singapore (1955) *Annual Report*, Singapore

assembly of goods for export elsewhere in the region, notably motor vehicles and bicycles, and since the 1930s there had been MNC manufacturing present in the shape of Fords and Bata. During the 1950s there was considerable expansion of manufacturing, mainly of consumer goods, for the local market. Employment in the manufacturing sector grew from 22,692 in 1955 to 44,295 in 1961. The most notable development was the establishment in 1953 of a spinning mill producing cotton, woollen and synthetic yarn. A significant feature of this undertaking was the recruitment of skilled labour from Hongkong to train local recruits (Fisher, 1964: 622). This was only one of a number of modern plants established on the Colonial Development Corporation's industrial estate at Bukit Timah.

To a degree similar patterns of development were taking place in the other port-cities of the region, but Singapore was by far the most advanced. This was a reflection of the opportunities and purchasing power generated by the entrepôt functions and British military-based facilities. Limited as the developments of the 1950s now appear, they gave Singapore the advantage of a nucleus of workers accustomed to factory employment.

By the late 1950s the Singapore economy, with 70 per cent of its GDP derived from entrepôt activities, was experiencing the impact of low commodity prices following the end of the Korean War. Manufacturing development was slow, having stagnated at 12 per cent of GDP, population was increasing at 4 per cent a year and unemployment rising, reaching 7–8 per cent in 1960. In addition political uncertainty (see chapter 4) and associated labour unrest were discouraging investment.

In 1961 a United Nations mission[4] produced a wide-ranging report on the economy and recommended that:

only manufacturing development would generate jobs at the required rate;
labour-force training and education programmes should be established;
there should be changes in labour relations to reduce unrest;
markets could be developed in the Federation of Malaya, UK and Europe;
four key industrial sectors be developed:
marine engineering
engineering
electrical equipment and appliances
chemicals;
overseas enterprise and investment should be attracted by the improvement of industrial relations, development of infrastructure and the establishment of incentive schemes;
an Investment Development Board and an Economic Development Board be set up to oversee the process of industrial development, provide investment where necessary and market Singapore overseas.

This report was to form the basis for the government development strategy, though in most instances, particularly with respect to labour relations, the policies followed were rather more far-reaching.

From 1961 until 1965 the strategy followed was strongly ISI, taking advantage from 1963 of the expanded market of the Federation of Malaysia. There was some increase in foreign investment, and the manufacturing sector's share of GDP increased sharply from 26 per cent in 1960 to 35 per cent in 1965. In addition, despite the emphasis on import substitution, the manufacturing sector's share of export earnings rose by 9 per cent.

The expulsion of Singapore from the Federation in 1965 seriously undermined the ISI strategy by dramatically reducing the size of the domestic market. The problems created by this were compounded by the announcement in 1967 of the closure of the British base. As well as providing for Singapore's defence the military installations were major sources of employment and revenue. In 1967 the naval dockyard alone employed 8,000 workers, the installations generating 16 per cent of total employment and 14 per cent of GDP (Buchanan, 1972: 86–7).

Fears were expressed by the government that the British withdrawal would leave Singapore with a limited defence capability and, in the political conditions prevailing in South East Asia, undermine the confidence of foreign investors. To counter this a series of diplomatic initiatives were taken to reduce tensions with Malaysia and Indonesia.

This process was greatly aided during 1967 by the installation of the pro-western Sukarto regime in Indonesia and the founding of ASEAN.

In 1967 conscription was introduced in Singapore and a rapid military build-up followed. The establishment of the 'garrison state' (see chapter 1) enabled Singapore to be presented during the late 1960s and early 1970s as a secure investment in a region that was viewed by investors as far from stable. In addition a large military capability gave the country additional prestige and was used in negotiations with the USA to present Singapore as a major bastion against 'communist subversion'. Perhaps more significantly, conscription was an important element of social control and a large proportion of defence expenditure was devoted to internal security.

From 1965 the development strategy moved sharply towards EOI. A wide-ranging development and promotional programme was initiated. This was largely based on the recommendations of the 1961 United Nations mission. Many of the policies operated against the interests of the indigenous manufacturing capitalist class that had moved into manufacturing. However, their activities had remained small in scale and they remained economically and politically weak (Rodan, 1987: 154). The ruling PAP had neither its origins or power base in this group (Pang, 1971). In this respect the Singapore situation contrasts strongly with other parts of the region.

From the early 1960s the PAP established control over organised labour and effectively stifled all opposition.[5] This was presented as a necessary step in the attraction of foreign investment and enterprise. The PAP launched a campaign of vilification against the trade-union movement, making constant reference to 'sabotage by the anti-national left':

> Take the case of another large firm producing electrical equipment. Here again, there is labour indiscipline, deliberately incited by their leaders who, of course, belong to another Barisan union.
>
> Again the production processes are jammed up by absenteeism. The mechanics deliberately set machines out of alignment whenever the union wants to bring pressure on the management. The culprits are members of the union committee and the company is unable to discipline them, as this would lead to a strike in the factory on the pretext of victimisation.
>
> Stories such as these come to the ears of overseas manufacturers and they naturally get frightened away. The Singapore worker, however, is both industrious and skilful. But he is being given a bad name through the lunacy of certain pro-Barisan trade union leaders. As a result of this, factories which would have gone up in Singapore are going elsewhere because the manufacturer has decided that he is not going to have the same unhappy experience.
>
> What are we going to do about this? It is clear that these pro-Barisan trade

union leaders are depriving countless young men and women of prospects of work through their sabotage of our industrialisation programme.

The government intends to set up a Commission of Inquiry to go into all these matters and expose the scandalous state of affairs in some of our factories. Intense public indignation at the misdeeds of these dangerous anti-social elements is certain, once the facts are known. So when the time comes to take action to curb them, the general public and workers in particular will say, 'And about time too.' (Goh Keng Swee, Minister of Finance, 1963: 12–13).

The Singapore Trade Union Congress was systematically undermined and in 1961 the basis of 'corporate trade unionism' was laid by the establishment of the NTUC (National Trade Union Congress). This, despite claims of independence, was a part of the PAP apparatus with successive general secretaries all holding cabinet rank (Miraz, 1986: 30). Thus balance between capital and labour interests was decisively shifted:

The Government already gives tax holiday for five years to pioneer industries. This means that they need not pay income tax on the profits earned during the five years. But this tax holiday is of no value to them if they are confronted with ceaseless rounds of wage demands and strikes.

So it is necessary that we arrange for wage stability for at least five years to allow these industries to find their feet and win in the competition with industrialists from other countries. This means there should be a wage agreement for a period of five years, during which time any dispute will be resolved in the first instance by negotiation, and if no agreement is reached, by reference to arbitration. (Goh Keng Swee, Minister of Finance, 1963: 16)

This shift was consolidated in 1968 by the Employment and Industrial Relations acts which abolished almost all workers' rights and established the ascendancy of the government and management. Since 1972 wage levels have been set by the NWC (National Wages Council).

The success of the PAP in 'taming the unions' is shown by the rapid decline in the number of disputes, days lost through industrial action and membership. While it is difficult to judge how important the image of a quiescent, organised labour force has been in attracting foreign investment and MNCs, the PAP has made much use of it in 'selling' Singapore. Indeed Ng Pock Too, Deputy Director of the NTUC and Political Secretary of the Prime Minister's Office, has informed potential investors that:

The labour movement in Singapore has always preferred the cooperative approach that can help employers facilitate flexibility, improve competitiveness and adapt swiftly to new technological development. (1986: 8)

The removal of trade union rights and the suppression of political opposition were accompanied by major housing and welfare programmes. During the mid-1960s these activities absorbed as much as 58 per cent of recurrent government expenditure. By 1988 urban renewal and township building had supplied 88 per cent of the population with public housing.

Since the mid-1960s major education and training programmes have been established. These have been closely related to the needs of domestic and, more especially, foreign manufacturing concerns. Four main areas have been developed: higher education, where the emphasis has been on the production of engineering graduates; vocational training, aimed primarily at school leavers; the Skills Development Fund which reimburses employers for sponsoring training and the updating of skills; and the training centres and units funded jointly by the Economic Development Board and various multinational corporations.

The totality of labour, welfare, education, training and housing programmes has produced by far the most literate, skilled and stable labour force in South East Asia. It is, however, by far the most heavily controlled. Few areas of life escape some form of regulation. The most notorious has been Lee Kuan Yew's incursion into eugenics. Better-educated women were exhorted to find partners of 'equal intelligence', while the less educated should have fewer children (*Far Eastern Economic Review Year Book*, 1987: 235; *Economist*, 1983: 50; Chee Chan Khoon and Chee Heng Leng, 1984). In 1983 the Social Development Unit was set up to help eligible graduates to meet!

As Mirza (1986: 269) has suggested, there is more than a hint of social-Darwinism about the authoritarian PAP regime. The major housing programmes were only one aspect of the PAP's improvement of basic infrastructure by heavy investment in power, water supply, sewerage, transport, post facilities and telecommunications. These efforts aimed principally at the creation of an attractive environment for foreign investment. The developments were implemented through a series of state monopolies such as the Public Utilities Board and the Telecommunications Authority of Singapore.

In order to promote key industrial sectors the state became involved directly in production through the establishment of a wide range of wholly and partly owned enterprises. These activities expanded rapidly. Many of the original concerns have diversified almost out of recognition. For example, the Keppel and Sembawarg Shipyards had, by 1981, grown into a broad-based group comprising thirty-four subsidiary and twenty associated companies, which included property-owning and financial services amongst their activities (Rodan, 1985: 176). The government has been prepared to give a lead in moving into new areas of production. This was particularly evident in the moves towards diversification from the early 1970s and into more capital-intensive activities since 1979. The latter is illustrated by the formation of SAMCO (Singapore Aerospace Maintenance Company).

The state has come to exert considerable control over the direction of domestic investments, in particular through the handling of the CPF (Central Provident Fund). Set up in 1955 as an employment insurance fund, in which the employer and employee each contributed 5 per cent of labour costs, to provide pensions, welfare payments and, later, the means to purchase housing, by the 1980s contributions had been successively increased to 50 per cent of labour costs (of which 27 per cent was paid by the employer). By 1984 the rate of saving was the equivalent of 42 per cent of GDP, the highest in the world. The equivalent figure for Japan was 31 per cent (Singapore Economic Committee, 1986a: 11). This extraordinary level of forced savings meant that the CPF had ceased to be primarily an insurance fund, but had become a major instrument of public finance (Harris, 1986: 62). With the addition of the Post Office Savings Bank the government has been in a position to direct the investment of the majority of domestic savings.

In 1961, following the recommendations of the United Nations, the Economic Development Board (EDB) was set up to coordinate research and promotional activity and take responsibility for the development of industrial estates. The expansion of these was rapid and in 1968 they were placed under a separate organisation, the Jurong Town Corporation. By the 1980s over 70 per cent of manufacturing employment was located on these estates. With some justification could Singapore be described as 'one big industrial estate'. Effectively, because of the island's 'free port status' the whole economy operated as an Export Processing Zone with no special privileges operating within the estates. There are, however, a whole range of incentives and concessions on offer for foreign enterprise.

The PAP's promotion of Singapore had a spectacular impact on foreign investment (figure 1.1). Investment by resident foreigners and foreign companies rose from 9 per cent of GDP in 1965 to 26 per cent in 1980. The majority of this investment went into manufacturing, where it represented 80 per cent of investment in 1982 (Miraz, 1986: 5). Between 1965 and 1980 the manufacturing sector was one of the fastest growing in the world, expanding from 1000 registered firms employing 47,000 people in 1965 to 3,300 employing 287,000 in 1980 (Smith *et al.*, 1985: 72). While only 13 per cent of these were foreign owned it has been estimated that they contributed 87 per cent of value added (Harris, 1986: 67).

Between 1965 and 1980 Singapore was one of the fastest growing

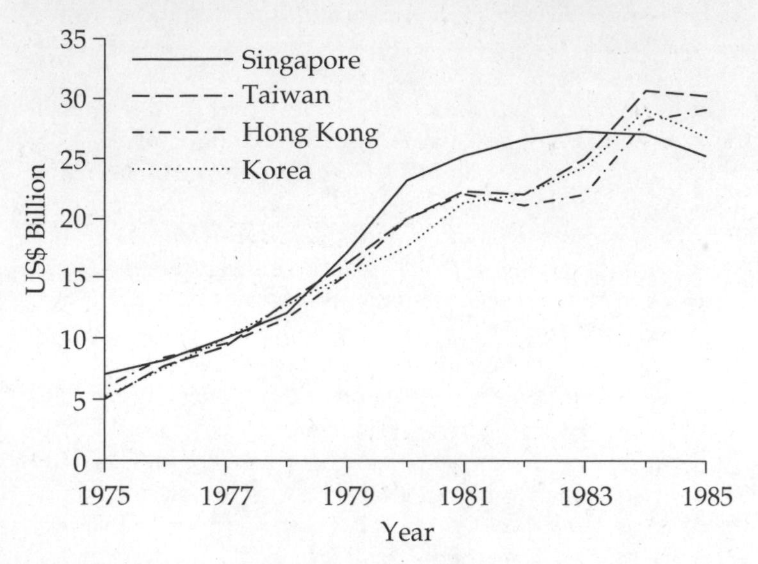

Figure 5.1 Exports of the Asian Newly Industrialising Countries.
Source: Singapore, Economic Committee (1986) *A review of Singapore's 1980–84 performance*, Singapore: 8

economies in the world with annual growth rates averaging 9 per cent. The expansion of export earnings was similarly spectacular, closely matching those of the East Asia NICs, Hongkong, South Korea and Taiwan (figure 5.1). However, the relationship between these growth rates and the manufacturing sector does merit closer inspection.

While the expansion of exports was impressive, the entrepôt function, particularly with respect to oil, exerted a major influence on the pattern of growth (figure 5.2). Indeed if re-exports and oil are discounted the growth of exports ceases to form a close match with those of the other Asian NICs, thus reflecting the degree to which Singapore remained closely tied to the regional trade in primary produce.

Despite the impressive rates of growth, Singapore has long run a sizeable balance of trade deficit. This has increasingly been met by the earnings of the service sector. In 1970 the service sector covered 34 per cent of the trade deficit and, by 1985, 97 per cent. From the early 1970s, in a deliberate move to diversify the economy further, Singapore was promoted as a major financial services centre. Essentially this involved building on the service sector which had grown up around the entrepôt activities and later broadened to serve the needs of MNCs. Singapore's 'location at the axis of ASEAN and its time-zone advantage bridging Europe, Asia and the Pacific makes it an obvious candidate for inter-

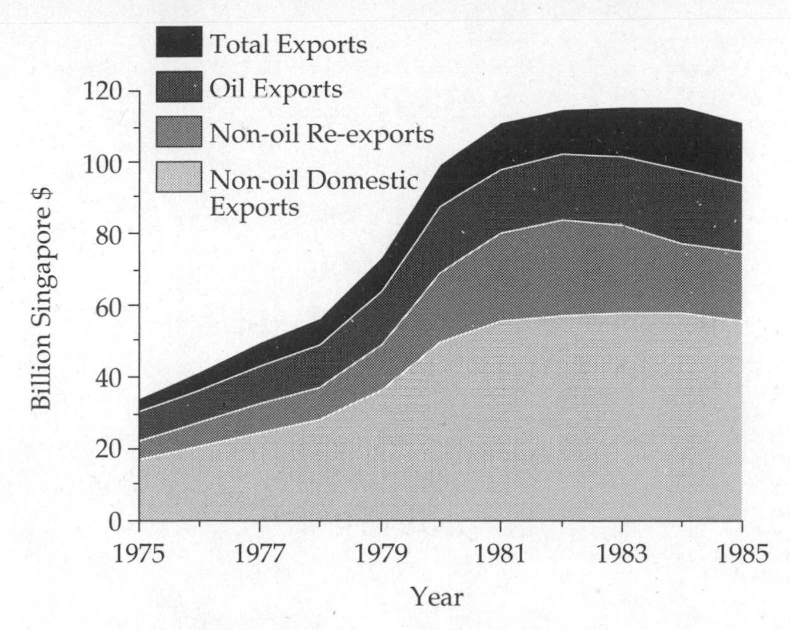

Figure 5.2 Composition of Singapore's exports, 1975–85
Source: United Nations (1987), *Year Book of International Trade Statistics*, New York

national banking seeking a regional base for operations' (Mirza, 1986: 123). These advantages were promoted vigorously by the Monetary Authority of Singapore (established in 1971).

Since 1968 the government has 'liberalised' the admissions policy to encourage the entry of international financial institutions. Tax rates have been reduced, for example, on currency transactions, and in some cases abolished, as on interest payments to non-residents. In 1978 all currency controls were abolished, thus all capital movements in or out of Singapore are unrestricted. These measures have given Singapore an advantage over Hongkong, which, for example, levies a 15 per cent withholding tax on interest paid to non-residents. While there is competition between Hongkong, Tokyo and Singapore there is also an increasing degree of complementarity (Hill, 1986: 6). Singapore has come to specialise in the Asian currency (see footnote to table 5.4) and Asian bond markets, while Hongkong has become the East Asian centre for fund management and loan syndication.

The financial service sector has grown very rapidly since the early 1970s and involves a wide variety of institutions, the vast majority of the new concerns being foreign owned (see table 5.4). Of the 108 foreign banks established by 1984, 30 were European, 27 American and 15

Table 5.4 *Commercial banks, Merchant banks and Asian Currency Units in Singapore, 1968–84*

	Commercial banks[1].						
		Foreign					
End of Period	Local	Total	Full	Restricted	Off-shore	Merchant Banks	Asian Currency Units[2]
1968	11	25	25	n/a	n/a	n/a	1
1969	11	25	25	n/a	n/a	n/a	11
1970	11	26	26	n/a	n/a	2	16
1971	11	31	25	6	n/a	2	21
1972	11	33	25	8	n/a	8	24
1973	11	43	24	12	7	17	46
1974	12	50	24	12	14	20	56
1975	13	57	24	12	21	21	66
1976	13	59	24	12	23	22	69
1977	13	64	24	13	27	23	78
1978	13	68	24	13	31	27	85
1979	13	76	24	13	39	33	101
1980	13	84	24	13	47	37	115
1981	13	95	24	13	58	41	131
1982	13	105	24	13	68	47	150
1983	13	109	24	14	71	51	159
1984	13	116	24	14	78	54	172

[1] The Monetary Authority of Singapore licenses three types of bank:
(a) full banks which are allowed to operate local branches and carry out the full range of banking activities;
(b) restricted banks which are not allowed to compete for local deposits;
(c) offshore banks geared to foreign operations.

[2] Asian Currency Units operate in the Asian Currency Market which is a short-term international money market equivalent to the Eurocurrency. It originated in Singapore in 1968 and is now located primarily there and in Hong Kong.

Source: Singapore Statistical News, 7, 3, 1985.

Japanese (Miraz, 1986: 133). Indeed the growth of the financial sector, like that of manufacturing, is predominantly a result of MNC activity.

In addition to the financial services Singapore has, since the early 1970s, offered incentives to a wider range of other 'producer services'.[6] Many of these, like the financial institutions, had originally been established to service MNCs and entrepôt activities. Barriers to foreign professionals practising in Singapore have been removed, and a variety of incentives introduced. In consequence there has been a rapid expansion of areas new to Singapore, such as technical services, data processing and marketing, while in established areas such as legal practice and accounting there has been increased foreign participation and sophistication of activities (Miraz, 1986: 154). As a result financial and business activities' share of service sector GDP rose from 39.7 per cent in 1960 to 68.7 per cent in 1983. Similarly in terms of service employment the share of finance and business increased from 6.7 per cent in 1960 to 12.9 in 1983 (Miraz, 1986: 30 and 179).

By 1978 Singapore was beginning to experience competition from lower cost labour locations and, at the same time, find expansion threatened by domestic labour shortage (Rodan, 1987: 157). According to McCue (1978) jobs were being created at a rate of about 40,000 a year while there were only 32,000–33,000 new entries to the labour market. This gap was increasingly met by the influx of foreign labour, particularly from Malaysia. There were two clear choices open to the PAP: either increase reliance on 'guest labour' or move towards more capital-intensive production. The government opted for the second strategy and in 1979 initiated a programme to encourage the development of higher valued added production and financial services. In the words of the Director of the Economic Development Board:

> We decided ... that making transistor radios was not a job for us. Nor do we want workers in the rag-trade. We want technical services. (cited Smith et al., 1985: 87)

There have been six main strands to the promotion of the 'second industrial revolution'. Firstly the inflow of foreign labour was curtailed and a 'wage correction' policy installed. Under this, wage increases of 54–8 per cent were recommended between 1979 and 1981. In essence this was aimed at forcing low-value, labour-intensive sectors to either upgrade or cease operation (Rodan, 1987: 159). Secondly, a range of tax incentives and cheap loans were offered to encourage higher value production. Thirdly, a number of Joint Industrial Training centres were established to increase the supply of skilled labour, particularly engineers. Fourthly, a major programme of infrastructure upgrading was announced, centring on the industrial estates. Fifthly, the PAP gave a lead, as it has done in the past, by direct investment and participation in production. The most significant of these developments was the petrochemical complex at Palau Ayer Marbau. Finally the PAP took steps to increase its already considerable control over the trade unions. In particular large groupings have been broken up and industry and plant level associations encouraged. Thus during the 1980s while the intervention of the state in the economy continued to follow the pattern established in the 1960s, it has reached a new level of intensity.

It is not easy to assess the success of the 'second industrial revolution' given the depressed state of the international economy during the early and mid-1980s. Between 1979 and 1982 foreign investment expanded rapidly (figure 1.1); however, the majority went into the service sector and construction. The growth of investment in plant and machinery declined from an annual rate of 23 per cent between 1978 and 1981 to 3 per cent between 1981 and 1984. (Singapore, Economic Committee 1986a: 12). There is, however, evidence of expansion of activities involving skilled labour in which Singapore retained a cost advantage. This was most clearly the case in electronics, where more capital-intensive and sophisticated production has developed. In 1984 two chip-making plants were established, one by the American-based Texas Instruments and the other by the Italian group SGS. This represented a major break in the dominance of the electronics sector by assembly work (see chapter 6 for a discussion of the significance of these developments in the context of the Asian Regional Division of Labour). However, the other major development between 1979 and 1982, the rapid growth of computer manufacture and the establishment of Singapore as a major exporter of disk-drives (*Asian Business*, December 1981: 46 and December 1982: 65), were essentially higher value assembly activities. These moves to higher value production characterised European and American investment but not Japanese since 1971. Japanese investment in Singapore has fallen; it has gone either to lower labour cost location or to Europe and the USA. This was a major blow to the PAP's strategy. The failure of international capital to respond fully to the PAP's 'second industrial revolution' reveals the limited extent to which even the most powerful Third World state economic apparatus can influence international capital.

The generally broader base of the economy to a degree insulated Singapore from the 1979–82 recession that so seriously affected the other NICs. However, sharp falls in commodity prices, including oil, slowed the growth of exports (figure 5.2) and made the Singapore economy increasingly vulnerable to any major contraction in the demand for its manufactured goods and services. This vulnerability is underlain by the fact that in 1984 66 per cent of goods and services were exported. The other Asian NICs are far less dependent on foreign demand – of their total production, Hongkong exports 48 per cent, Taiwan 37 per cent and South Korea 26 per cent (Singapore, Economic Committee, 1986a: 7).

In 1985 all the NICs experienced problems, but Singapore was, because of its more exposed position, by far the hardest hit. The fall in Singapore's industrial production was much steeper than in the other Asian NICs (figure 5.3). The impact of the sharp contraction of markets for Singapore's products was reinforced by the collapse of the domestic construction industry boom which between 1982 and 1985 was a major

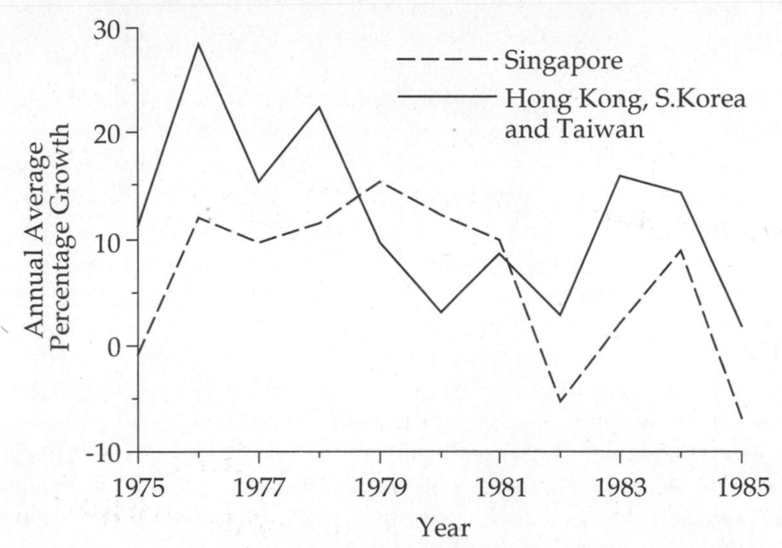

Figure 5.3 Percentage changes in the industrial production of Singapore and the other East Asian Newly Industrialising Countries.
Source: Singapore, Economic Committee (1986) *A review of Singapore's 1980–84 performance*, Singapore: 12

factor in sustaining growth rates.[7] Growth came to a shuddering halt. GDP decreased by 1.8 per cent, declines of 8 per cent in manufacturing and 17 per cent in construction being partly balanced by a 5 per cent growth in the service sector. There were widespread business failures and sharp falls in export earnings and, more spectacularly, foreign investment (figure 1.1). Unemployment rose from 2.9 per cent in 1984 to 6 per cent in 1986. In addition the number of Malaysian 'guest workers' fell from *c*. 105,000 to *c*. 60,000 (Sieh, 1988: 104).

The serious impact that the global recession had on Singapore, particularly between 1984 and 1986, reveals the vulnerability of such an open economy. These experiences resulted in some revision of the PAP's development strategies. These centred on the recognition that the leading growth area will in future be the service sector with manufacturing playing a major but secondary role (Singapore, Economic Committee, 1986a). In effect the government merely recognised what had been happening since 1980 and incorporated it into strategy. Singapore was to be promoted:

as a premier Operational Headquarters (OHQS) for Manufacturing, Finance and Banking Services, especially risk management and fund management, knowledge intensive industries and other internationally traded services. (Singapore, Report of the Fiscal and Financial Policy Sub-committee, 1986: 1)

While the aim of attracting high-value industry was retained, the importance of labour-intensive industry was also recognised. To this end the continual use of 'guest labour' was accepted. However, no major influx of foreign labour is contemplated and given the almost zero population growth labour supply is likely to exert a major constraint on the economy. Perhaps more significant was the realisation that Singapore has, since the late 1970s, lost the 'competitive edge' in manufacturing.

Between 1979 and 1984 total labour costs rose by 10.1 per cent while productivity per worker rose by only 4.6 per cent (Cass, 1985). During this period Singapore's competitive position declined by 50 per cent against Hongkong, 15 per cent against Taiwan and 35 per cent against South Korea (Singapore, Economic Committee, 1986a: 31). While rising labour costs were seen as the principal cause of this decline, attention has also been drawn to the role of increased rents, interest charges, transport costs and the costs of power and water in reducing the level of profitability (Singapore, Economic Committee, 1986a: 44).

A variety of measures has been introduced with the collective aims of reducing production costs, raising profitability and attracting foreign investment. Taxation on charges for electricity, gas and telephone calls was suspended and 50 per cent rebates on property taxes granted. Particular attention has focused on the reduction of corporation tax from 40 to 33 per cent (*Far Eastern Economic Review Year Book*, 1987: 237). However, central to the policy has been the reduction of labour costs, through cutting of the employers' contribution to the CPF from 25 to 10 per cent, wage freezes and actual reductions. These policies are being pursued with the active 'cooperation' of the corporate trade unions.

It is clear that since 1985 there has been a significant retreat from the 'second industrial revolution' strategy (Rodan, 1987: 172). Whether the revised policies will enable Singapore to regain the 'competitive edge' is a moot point. The government's ability to reduce labour costs far enough to compete with lower cost areas is, in particular, highly debatable. Since 1985 the government has attempted to stimulate the economy by increasing development expenditure. By 1986 state expenditure as a percentage of GDP had risen to 29.3 per cent (table 5.5). Given the reductions in tax and CPF contributions this is only sustainable if major economies are made in recurrent expenditure and a high level of capital inflow is established.

Table 5.5 *Government expenditure as a percentage of GDP, 1965–87*

	Indonesia	Malaysia[1]	Philippines[1]	Singapore	Thailand
1965	n/a	n/a	9.7	n/a	16.1
1966	n/a	24.3	10.2	20.4	16.7
1967	n/a	24.6	10.5	19.0	18.2
1968	n/a	25.6	11.3	18.3	18.4
1969	12.3	22.7	11.6	19.0	18.5
1970	13.7	24.7	10.4	21.5	19.8
1971	14.8	29.6	10.9	22.3	19.9
1972	16.1	33.9	13.5	20.5	18.5
1973	17.2	27.5	17.5	23.1	16.0
1974	18.4	31.4	18.3	20.4	16.2
1975	22.4	34.4	18.7	23.7	18.4
1976	23.8	30.4	20.0	23.0	19.0
1977	18.6	27.8	13.3	20.0	16.9
1978	19.4	25.2	13.2	18.9	17.8
1979	20.6	22.7	11.7	18.6	17.9
1980	22.3	28.5	12.3	19.2	19.1
1981	24.1	38.4	12.7	22.7	19.0
1982	20.9	36.1	12.0	20.7	20.4
1983	21.2	31.2	11.7	22.0	19.9
1984	18.9	27.6	9.8	26.3	19.6
1985	21.9	n/a	10.5	27.3	21.1
1986	21.5	34.4	13.0	29.3	20.4
1987	22.1	29.0	n/a	n/a	19.2

[1] 1965–76 figures are as a percentage of GNP
Source: 1977–87, IMF *Government Financial Statistics Year Book*, XII, 1988, Washington; 1965–76, *Asian Business Quarterly*, 3 (4), 1979: 46

Concern over the level of public expenditure and the need for the state to revitalise the economy by promoting new areas of activity has, since 1986, led to a systematic policy of privatisation. The Public Sector Disinvestment Committee was established to formulate a programme of disinvestment in Government Linked Companies (GLCs). Some 634 of these concerns together with 7 of the 33 Statutory Boards have been considered. However, while substantial progress has been made the state has retained a 51 per cent controlling interest (Ng Chee Yuen 1989). In addition, investment has brought new areas of the economy under central government control. Thus assets are being realised in some sectors of the economy in order to invest in others. Ng Chee Yuen and Wagner (1989: 216) have suggested that there is unlikely to be any significant reduction in the involvement of the state in the economy; rather the outcome 'may be just a change of company portfolios and activities; and public enterprise would continue to be the instrument of restructuring and development'.

The various policies implemented since 1985 have contributed to a restoration of confidence on the part of domestic and, more especially, international investors. However, given the highly internationalised nature of the Singapore economy – in 1988, 88 per cent of total demand was external (Balakrishnan, 1989: 66) – the renewed levels of growth that the city state has experienced (table 1.19) are primarily a reflection of a partial recovery of the world-economy. In particular, renewed expansion in the demand for electronics resulted in rapid growth in this industry and the sectors that underpin it. Indeed it has been the manufacturing sector that has been the main element in Singapore's recovery rather than finance and business services. However, during 1989 the emergence of a world surplus of disk-drives heralded a contraction in the electronics sector and a marked slowing of rate of growth of Singapore's manufacturing sector and economy as a whole.

In addition to Singapore's vulnerability in a period of marked instability in the world-economy the totality of government policies since 1981 has caused increasingly widespread discontent and focused opposition to the PAP. Such opposition may well limit the government's ability to follow present strategy as well as undermining it by reducing the confidence of foreign investors.

Thailand

In many ways Thailand is the regional polar opposite of Singapore. While manufactured exports expanded rapidly from the early 1970s the country has never made a clean break from ISI. In addition until the 1980s the economy was probably the least internationalised of the ASEAN groups with a low level of foreign investment (tables 1.6 and 1.7) and MNC activity. During the immediate post-war period Thailand was probably the most 'open' of the South East Asian economies and the one in which capitalist production had made least headway. However, these conditions did not result in any major influx of foreign activity. The lack of any resource not readily available elsewhere in the region, location off the main regional and international transport routes, apparent instability of the governments and proximity to the former Indo-Chinese states combined to discourage foreign undertakings and investors. In addition, particularly between 1947 and 1957, the nationalistic and often erratic nature of government economy policy tended to discourage investors.

In the immediate post-World War Two period Thai economic development was characterised, like most of South East Asia, by intense nationalism. During this period the Chinese population was heavily discriminated against. The government of Phibun Songkram (1947–57) established a form of state-led capitalist development. Under this the interests of Chinese and foreign capital were secondary to those of Thai nationals (Skinner, 1958: 317; Ingram, 1971: 287, 275). Foreign investment and manufacturing activity were inhibited by threats of closure, state-promoted competition and non-renewal of contracts.

During the period 1947–57 development strategy, while never clearly stated, had all the characteristics of ISI. However, these policies did not centre on tariff protection. Tariffs were treated primarily as sources of revenue. In addition, Akrasanee and Juanjai (1986: 77) have suggested that the Phibun regime 'deliberately avoided protecting industries for fear of promoting the Chinese community'. Rather, industrialisation was promoted through the direct participation of the state in production.

State enterprises were set up with monopolies in such areas as brewing, paper manufacture, sugar refining and gunny sack production (Marzouk, 1972: 195; Silcock, 1967: 262–3). While some were fully state-owned, others were under the control of NEDCOL (National Economic Development Corporation Limited). The majority of these undertakings were effectively a partnership between the state and major Thai and Chinese business interests (Hewison, 1985: 276; Ingram, 1971: 288). Through NEDCOL the state acted as a guarantor of loans enabling comparatively large-scale foreign funding to be obtained for many of these enterprises. However, direct state investment played a major role in the development of manufacturing. The government's share of domestic investment rose from 32 per cent in 1952 to over 38 per cent for the period 1953–5 (Akrasanee and Juanjai, 1986: 83).

Evaluation of the state enterprise programme is difficult. The state concerns became bywords for corruption and inefficiency, none of them ever reaching target production levels (Ingram, 1971: 287). In addition, after the overthrow of Phibun in 1957, the record of NEDCOL and the state enterprises became a favourite area of propaganda to further discredit his regime and the direct involvement of the state in production. However, the state enterprises, whatever their failings, did initiate domestic production in a number of important areas and set Thailand on the ISI road.

By 1957 there was mounting opposition to the Phibun development strategy. This stemmed from domestic and foreign investors, the World Bank and the Bank of Thailand. The protests do not appear to have been directed at ISI *per se* but rather at the haphazard way in which the state intervened in the economy. In particular, major domestic business interests were lining up against the Phibun regime. Many of these represented large amounts of capital that had been accumulated by trading activities during the Korean War which was unable to find domestic outlets under prevailing government policy and the declining trade in primary produce. Hewison (1985: 276) has shown how investors were inhibited because they could never be certain that the state intervention 'in the national interest'[8] would not take place in their chosen field.

Thus the Phibun regime was seen as a major barrier to the diversification of the economy and the activities of foreign and domestic capital. In addition, development was being seriously hindered by poor infrastructure and a lack of any coordinated planning machinery. These deficiencies were highlighted by the report of the IBRD 1957–8 mission.

The end of the Phibun government and the installation of Field Marshall Sarit as Prime Minister in 1959 marked a decisive change in development strategy. Sarit and his backers were strongly opposed to direct state involvement in production. Thus policy shifted towards the development of infrastructure, administration and planning machinery which would create an environment suited to the long-term expansion of domestic and foreign investment.

The recommendation of the 1957–8 IBRD mission, that coordinated national planning should be established, was reinforced by fears that American grants and soft loans would be withdrawn (Caldwell, 1974: 49), concern over the long-term prospects for economic growth and the perceived need to direct development to impoverished areas of the kingdom where insurgency was increasing. Thus in 1960 the NEDB (National Economic Development Board) was established, and immediately began the formulation of the First National Development Plan (1961–6).

The First National Plan contained a significant regional element that was directed in particular towards the North East of the kingdom. The post-war recovery of the Thai economy brought the North East's poverty into focus. This was considerably sharpened by what was seen as 'deteriorating security' within the region and the events in neighbouring Indo-China (Nakahara and Witton, 1971: 52). North Eastern Members of the National Assembly were becoming increasingly vocal about the poverty and neglect of their region (Keyes, 1965). As a result, in the 1950s piecemeal development schemes aimed at transport, health, education, water supply and irrigation were established. Under the First Plan the development of the region became coordinated by the North Eastern Economic Development Sub-committee. Under the first (1961–6) and second (1967–71) National Development Plans the North East absorbed 28 per cent of the total budget, a percentage only slightly less than its share of the population.

Between 1961 and 1971 policy for the North East stressed the development of infrastructure, particularly transport and irrigation, and to a lesser extent health, welfare, education and agriculture (Pakkasem, 1973: 15–16). During this period Thailand received a significant volume of military and other 'aid', principally from the United States. While American economic 'aid' has been used for a wide variety of purposes in the kingdom as a whole, the most important of which have been roads, health, education and police administration, increasingly the primary emphasis was placed on accelerated rural development schemes in the North East. The American military presence in Thailand grew rapidly during the 1960s, and major bases for operations in Indo-China were established in the North East (Peagram, 1976: 10–12). These gave rise to considerable, if localised, transport, urban and service-industry developments. Additionally American military personnel were involved in the North Eastern Communist Suppression Command which became increasingly active during the late 1960s when 'insurgency' was reported for twelve of the sixteen North Eastern provinces (Nakahara and Witton, 1971: 56). Rural development was seen as of major importance in maintaining and increasing loyalty to the government. As a Thai Deputy Prime Minister put it: 'If stomachs are full people do not turn to communism' (Air Chief Marshal Dawee Chulasapp, quoted in the *Bangkok World*, 17 November 1966). A series of Thai–American 'rural development' schemes were initiated from the mid-1960s. These combined rural development with propaganda stressing the evils of communism and the benefits that would stem from the government (Caldwell, 1974: 135).

While the main emphasis of the 1961–6 plan was on the rural sector and the expansion of agricultural production, the development of the manufacturing sector received considerable importance. To this end there was heavy investment in infrastructure, particularly transport and power generation (50.9 per cent of the plan budget). Industrial development was promoted by the introduction of a range of tax incentives and the establishment of selective tariff protection.[9] The promotional measures were coordinated by the BOI (Board of Investment established in 1959); they were broadly similar to those established elsewhere in the region.

From the early 1960s national and manufacturing growth rates accelerated. The degree to which this was due to the promotional programme and government policy in general remains uncertain. However, the increased tariff protection undoubtedly provided some encouragement for manufacturing investment (Sukwankiri, 1970). More significantly, the demand for Thai primary produce was buoyant, and the expansion of domestic buying power combined with comparative political stability encouraged Thai, Chinese and overseas investment. In addition, the escalation of the Vietnam war brought increasing amounts of American military and economic 'aid' flowing into the country (Caldwell, 1974; Elliott, 1978: 131).[10]

While foreign investment grew rapidly during the 1960s its share of total investment fell as domestic capital moved increasingly into what were regarded as 'safe areas'. Despite this the 1960s were formative years in the development of a multi-national presence in Thailand. According to Akira Suehiro (1985, cited Hewison, 1987: 58) perhaps half the MNCs operating in the country during the 1980s were established there between 1963 and 1972.

During the 1960s the basis of an import substitution sector was laid in Thailand. Ingram (1971, 297–8), in a detailed study of the impact of ISI during the 1960s, concluded that with limited exceptions:

> the primary effect of investment promotion in Thailand has not been a substitution of domestic production for imports, but a substitution of one kind of imports for another – in particular, a substitution of imported raw materials, components and capital goods for imports of the products now turned out by promoted firms.

In this respect Thailand's ISI experience is similar to that of the other cases described in the chapter and indeed of the Third World in general.

There was a tendency for the promotional concessions to be renewed after the initial five years, for protective tariffs to rise (until the early 1980s) and the duties on imported inputs to fall. As elsewhere the development of the sector was inhibited by the small size of the domestic market. The majority of products were aimed at the 'urban enclave market rather than the mass rural market' (Richter and Edwards 1973: 38). This was reinforced by the tendency of the BOI to attempt to prevent promotional firms establishing monopolies by splitting protection between several firms (Ingram, 1971: 297). During the late 1960s the domestic consumer market, particularly for textiles and locally assembled vehicles, was at near saturation point and the rate of growth of the manufacturing sector was faltering.

By 1970 the Thai government was beginning to reassess its commitment to ISI in the light of growing financial problems. The slow down of the world-economy and the reduction of the American economic and military presence[11] were presenting Thailand with widening balance of payments and budget deficits. While exports had been diversified since the late 1950s, the country remained heavily dependent on primary products (table 1.12). In 1971 the BOI announced incentives for export promotion, which were enhanced by the 1972 Investment Promotion Act; the Third Plan, 1972–6, emphasised both ISI *and* EOI.

By the mid-1970s there was increasing pressure for a further shift in emphasis towards EOI. This was coming from the international agencies, particularly the World Bank (1980a), the need to increase exports to reduce the balance of payments deficit and domestic capital interests. These influences were strongly reflected in the Fourth Plan, 1977–81, under which EOI was given a central role. However, the continued importance of ISI was also stressed.

In 1977 and 1979 incentives for export-oriented industries were increased. Additionally, in 1978, Thailand followed the example of South Korea and Japan by establishing thirteen International Trading Companies. These were intended to channel exports into new markets. In the absence of these their success was limited; by 1984 only 4 per cent of manufactured exports passed through them (*Bangkok Bank Review*, 25, 7, 1984: 253–8).

During the Fourth Plan attempts were also made to promote industrial estates. Proposals for estate development had been drawn up by the BOI as early as 1961 but nothing was constructed until 1971 when a small, 114 hectare estate was constructed at Bang Chan. In 1972 the Industrial Estates Authority of Thailand (IEAT) was established and two further estates developed at Lat Krabang in 1975 and Bang Pu in 1977. In 1979 a major programme of estates expansion was announced and this was given considerable prominence under the Fifth, 1981–6 Plan. In 1980, an EPZ was established at Lat Krabang, involving a major expansion of the original industrial estate (IEAT, 1983). The economic problems of the 1980s have seriously curtailed all these developments, only four of the eleven planned estates having been implemented. Despite the availability of additional incentives (*Bangkok Bank Review*, 24, 12, 1983: 495) little investment has been attracted.

When viewed in total the promotional programme has, over the last 25 years, been far from impressive. Between 1959 and 1985 1,685 firms were issued with certificates. Of these only 1,037 companies have actually started operations providing a total investment of US$ 5,729.8 million. Only 10 per cent of these were entirely foreign-owned concerns, though a further 23.4 per cent comprised joint Thai–overseas undertakings. The vast majority of promotional investment – 73.4 per cent – has been Thai. However, the pattern of ownership in the manufacturing and financial sectors is far from simple and by no means easy to identify (Hewison, 1985: 273–4). For, as in the Philippines, much of the capital for foreign-owned concerns is raised locally.[12] Table 5.6 provides a general indication of the degree of foreign ownership. The high level of

Table 5.6(a) *Foreign Ownership of the Thai financial sector, 1979*

Financial activity	No. of foreign companies	Assets (in millions of baht)	% of market controlled
1 Commercial banks	14	18,106	6.15
2 Life insurance	3	1,282	32.10
3 Non-life insurance	9	447	16.15
4 Investment and securities[1]	15	13,797	27.58

[1] Includes joint ventures and foreign companies registered in Thailand
Source: Hewison, 1985: 274

Table 5.6(b) *Ownership of various sectors of Thai industry, 1978–79*[1]

Sector	% foreign-owned	% Thai-owned
1 Auto industry[2]	55.5	44.5
2 Textiles[3]	54.7	45.3
3 Petroleum	96.0	4.0
4 Cement	insignificant	100.0
5 Sugar	2.0–10.0	90.0–98.0
6 Tin	90.0	10.0
7 Steel	20.0	80.0
8 Milk	60.0–70.0	30.0–40.0
9 Soap products	65.0	35.0

[1] Unless otherwise stated based on total assets of companies
[2] Based on registered capital data reported in *Business in Thailand*, May, 1980: 112–13
[3] Based on registered capital data reported in *Business in Thailand*, August, 1976: 61–4
Source: Hewison, 1985: 274

foreign ownership in the long-established tin sector and the newly developed oil industry is particularly striking.

While the manufacturing sector and production in general remained predominantly under domestic control, the inflow of foreign funds has increased in value and share of total capital. While these inflows were in part due to balance of payments and government budget deficits, a considerable volume of funds did become available to local and foreign investors through local financial institutions. Whereas in the Philippines the process operated in addition to a high level of direct foreign investment, in Thailand it appears to have operated in place of it.

Further, as Hewison (1987: 60) has noted 'the supply of large loans and credits by international agencies and private transnational banks gives them a considerable stake in the direction of national development and provides international finance capital with a strong stimulus to influence development policies'. Thailand's ability to borrow overseas, despite the comparatively low levels of foreign investment provided a major cushion during the 1970s. This was reinforced by the country's, in regional terms, comparatively low degree of integration in the world-economy (table 1.10). Despite a series of short-term crises the economy maintained comparatively high growth rates (tables 1.3 and 1.19) and staved off the main impact of the recession until the 1980s. However, the inflow of funds had its penalty in a rapidly escalating overseas debt (table 1.18).

As was outlined in chapter one, during the 1970s Thailand, like other oil-importing primary producers experienced recurrent economic crisis and generally widening balance of payments deficits. The sharp fall in commodity prices during the period 1979–81, coupled with the rise in oil prices, brought the economy to the verge of collapse. As a result during the early 1980s Thai economic policy became closely supervised by the IMF and IBRD.

In late 1980 the World Bank produced an economic report on Thailand and recommended a five-year programme of 'structural readjustment'. The main policy recommendations were: to raise domestic energy prices to the international level; the development of a strong deflationary monetary and fiscal policy; an end to the import substitution policy for industry, and emphasis to be placed on export-oriented industries; reduction of import tariffs and the removal of all export restrictions and taxes; increased and more effective personal taxation; an end to restrictions on the level of domestic interest rates; and a comprehensive review of government organisation and expenditure in order to eliminate waste. These recommendations were, early in 1982, made the conditions for a $150m SAL ('Structural Adjustment Loan'). These reforms were adopted as integral to the Fifth Plan (1982–6). Their implementation was also a condition of Thailand's receiving SALs in 1982 and 1983, totalling $225m. Progress was very limited; in 1984 and 1985 targets were substantially reduced, and plans for further SALs were abandoned.

In mid-1985 Thailand approached the IMF for funding, reaching agreement on a two-year programme of support, totalling $586.6m. The programme was designed to mitigate the effects of the continued deterioration of the balance of payments. Government statements minimised the significance of the conditions that were attached to the agreement, to avoid creating the impression among foreign banks that the country was on the verge of a debt crisis, thereby reducing Thailand's favourable international credit rating. Similarly, for domestic reasons, the government wished to avoid any public concern that the loan would increase economic hardship.

While the Thai government has been reluctant to fully implement the structural adjustment proposals, development policy has generally moved in that direction. Promotional procedures have been simplified and restrictions on foreign concerns, particularly in the financial sector, eased. Government expenditure has moved away from regional and rural development and firmly towards the creation of an environment suitable for major industrial expansion.

The Sixth Plan (1987–91) is chiefly directed at the stabilisation of the economy. It seems likely that this aim will take precedence over economic growth. However, government debates suggest that there is still strong support for a policy of growth maximisation. The equalisation of regional and personal income levels are expected to receive even lower priority than under the Fifth Plan.

One of the main objectives of Thai development policy in the 1980s is the establishment of a major industrial complex on the Eastern Seaboard, based on the exploitation of offshore gas and oil. The hope that the gas developments, with low Thai labour costs, will attract the necessary foreign investment has not yet been realised, and an increasing number of the planned projects are being postponed, cancelled or scaled down.

The economic problems of the 1980s have brought into the open the conflicts between ISI and EOI interests. Despite the large number of changes that have taken place in export promotion policies since 1972 the distinction between ISI and EOI has never been satisfactorily clarified. High levels of tariff protection remain despite the pressure exerted by the World Bank and IMF and conditions of the SALs. Indeed in many areas, for example the assembly of motor vehicles, controls have increased in an effort to reduce the dependence on imported components.

Despite the many shortcomings and recurrent problems, the Thai manufacturing sector has shown remarkable development since the mid-1960s (table 1.14). However, the sector remains dominated by small-scale, family-controlled plants, predominantly producing for the domestic market. A small number of major industrial groups has emerged, most notably the broadly based Siam City Cement, the textile-based Saha Union Group and the agro-industrial Charen Pokyhand Group. Since the mid-1970s these concerns have been increasingly orientated towards the export market. The manufacturing sector remains heavily dependent on imported raw materials, components and capital goods (table 5.7). With the major exception of the agro-processing areas the development of manufacturing that utilised domestic raw materials has been limited. In 1972, for example, only 6,000 tons of

Table 5.7 *Thailand, changing composition of non-energy imports, percentage of total values*

	Consumer goods	Intermediate goods and raw materials	Vehicles and parts	Capital goods	Total
1965	29.5	21.8	10.1	32.6	55.5
1970	14.1	27.2	9.9	37.9	75.0
1975	16.8	31.2	8.5	41.8	81.5
1980	14.8	34.8	5.6	35.1	75.5
1984	17.0	32.8	6.2	38.5	77.5

Source: Bank of Thailand, *Monthly Review*, various issues

local rubber (1.8 per cent of production) was used locally; despite efforts by the BOI in 1985 this had only risen to 30,000 tons (5.6 per cent of production) and Thailand remained a major importer of rubber goods.

The development of EOI in Thailand has not 'internationalised' the economy in the sense of the manufacturing sector becoming dominated by foreign investment and multinational corporations. Despite its remarkably high international credit rating and, in the 1980s, the much-vaunted label of 'the dark horse of the ASEAN economies' (Bowring, 1982: 85; Dixon, 1984: 66), it has remained very much on the fringes of this type of development. Political uncertainty and rising comparative labour costs elsewhere have, since 1983, increased the MNC presence, particularly in electronics, and largely at the expense of other parts of the region. However, this new cycle of capitalist penetration was, by 1988, beginning to falter, largely because of the poor infrastructure provision, limited power generation and shortage of workers skilled in manufacturing processes (Handley, 1988: 94–5).

Malaysia

Malaysian development strategy has been heavily influenced by the colonial legacies of uneven development, plural society and lack of national integration. While, as was discussed in chapter 4, these features were common to the entire region, they were at their most all-pervading in Malaysia. Most striking was the marked 'ethnic division of labour' that had developed during the colonial period (table 5.8). The Malay population was very largely excluded from cashcrop production and the urban-industrial sector. While immediate post-independence development policy set out to protect and promote the interests of the

Table 5.8(a) *Peninsular Malaysia: employment by occupation and race, 1970*

Race	Primary	Secondary (includes mining)	Tertiary	All sectors
Malay	67.6	30.8	37.9	51.4
Chinese	21.4	59.5	48.3	37.0
Indian	10.1	9.2	12.6	10.7
Other	0.9	0.5	1.2	0.9
Total	100.0	100.0	100.0	100.0

Source: Malaysia, 1976: 181–2, cited Majid and Majid, 1983: 68

Table 5.8(b) *Peninsular Malaysia: mean household incomes, 1970*

Groups	Mean household income ($ per month)	Poor households (%)
Malay	172	74
Chinese	394	17
Indian	304	8
Others	813	1
Rural average	200	86
Urban average	428	14
All households	264	

Source: Malaysia, 1976: 179, cited Majid and Majid, 1983: 69

Table 5.8(c) *Peninsular Malaysia: urban residence by ethnic group, 1970*

Group	%
Malay	27.6
Chinese	58.5
Indian	12.8
Other	1.1

Source: Malaysia 1970, *Population and Housing census*, Dept. of Statistics

bumiputeras[13] there was considerable inter-racial tension. This came to a head in 1969 with the so-called '13th May' riots. Behind these disturbances lay very major differences in the incidence of poverty and access to financial opportunities. Following these disturbances the NEP (New Economic Policy) was adopted which centred on accelerated economic growth and the promotion of the interests of the *bumiputeras*.

The NEP was a long-term socio-economic policy,

designed to achieve national unity through the two-pronged objectives of eradicating poverty irrespective of race and restructuring society, to eliminate the identification of race with economic function. (Malaysia, 1976, Third Malaysian Plan 1976–80: 2)

In practice this meant that the incomes of the predominantly rural Malays were to be raised through a variety of development, community and land-colonisation programmes. In addition, a number of state enterprises and statutory bodies were established with the intention of increasing Malay participation in commercial and industrial activities (Wong, 1979: 64). The long-term objective, as stated in the 1971–5 plan was that 30 per cent of commerce and industry would be under Malay control by 1990.

These new policies involved a high degree of state economic intervention. Government expenditure increased sharply and came to absorb by far the largest share of GDP of any South East Asian economy (table 5.5).

In 1957 the Malaysia economy was one of the most narrowly based in South East Asia, with 85 per cent of export earnings derived from tin and rubber. Some 70 per cent of the profits from registered companies were in foreign, mainly British hands and were largely repatriated. Agency houses, largely European owned, controlled 70 per cent of foreign trade and 75 per cent of plantations (Lee, 1986: 102). However, almost 50 per cent of accumulated capital was locally owned, predominantly by Chinese (Gill, 1985: 5). Thus a considerable proportion of the profits from the 1950s commodity boom remained within the country and played a major role in the post-independence expansion and diversification of the economy (Thoburn, 1977).

Independence brought no significant reduction in the degree of foreign control over the economy. Malaysia continued to be regarded as one of the most open, stable and attractive areas for investment, particularly in primary production, in South East Asia. During the 1960s the foreign-owned plantation companies invested heavily in diversification out of rubber and into tobacco, cocoa and, more especially, oil palm[14]. This process was furthered in the early 1970s by the development of oil production. In addition, the predominantly British-owned agency houses began to invest in import substituting industries (Sundaram, 1987: 113–14). Thus the Malaysian economy, while remaining closely tied to foreign interests, underwent rapid expansion and diversification.

While considerable emphasis was placed on industrial and infrastructural development agriculture was initially earmarked as a priority area (table 5.9). This reflected both the importance of the agricultural

Table 5.9 *Malaysia development expenditure 1956–90, percentage distribution*

Expenditure category	1956–60	1961–65	1966–70	1971–75	1976–80	1981–85	1986–90
Agriculture	23	18	25	22	21	12	18
Rubber replanting	15	5	3	1	1	1	3
Drainage and irrigation	4	4	9	2	3	2	[1]
Land development	2	5	9	11	11	4	6
Other	2	4	4	8	6	5	8
Transportation	23	22	10	18	14	17	16
Road	12	16	5	11	9	13	11
Rail	7	2	1	1	1	1	2
Port	4	2	2	4	3	2	–
Aviation	[1]	2	1	2	1	1	–
Communication	5	4	4	6	4	7	14
Telecommunication	5	3	3	5	4	7	11
Broadcasting	[1]	1	1	1	[1]	[1]	[1]
Post	[1]	[1]	[1]	[1]	[1]	[1]	[1]
Utilities	24	24	18	10	7	12	14
Electricity	14	13	14	6	5	9	9
Water supply	8	7	4	3	2	2	4
Sewerage	2	[1]	[1]	[1]	[1]	[1]	1
Commerce and industry	1	2	4	17	14	27	14
Social services	14	16	18	13	17	13	13
Education and training	6	9	8	7	7	6	8
Health and family planning	1	4	3	2	2	1	1
Housing	6	3	5	2	5	5	3
Social welfare and other	–	[1]	2	2	3	1	1
General administration	6	6	3	3	3	1	4
Defence	3	9	15	7	16	8	4
Internal security	1	2	3	3	4	2	3

[1] indicates less than 0.5 per cent
Source: Malaysian Development Plan, 1965, 1971, 1976, 1981 and 1986, compiled by Cho (1990)

exports and the need to reduce the dependence on rice imports. In 1960 35–40 per cent of rice was imported. Thus the government encouraged the introduction of new export crops and the expansion of rice production. The emphasis on the agricultural sector, while perhaps reflecting a realistic appraisal of the position Malaysia occupied in the world-economy, also reflected the political representation of rural Malay interests institutionalised in the NEP after 1969 (Rudner, 1975).

A wide range of public organisations have become involved in rural development. These major bodies, FELDA (Federal Land Development Authority), FELCRA (Federal Land Consolidation and Rehabilitation Authority) and RISDA (Rubber Industry Smallholders Development Authority), are all principally orientated towards land resettlement. In addition, there are a multitude of other agencies operating in rural areas. However, the efforts of these various organisations are far from fully coordinated and in many cases functions and activities overlap if not conflict with one another (Majid and Majid, 1983: 94).

Despite the emphasis placed in planning documents and government statements on the importance of raising the incomes of the rural Malays, considerable doubts have been cast on the effectiveness of policies. Resources were largely channelled towards the landed and cash-crop interests and not into raising the incomes of the small Malay rice farmer. Rural communities have become increasingly polarised (Gibbons *et al.*, 1986; Scott, 1984) and there is little evidence of widespread reduction in the incidence of poverty (see section 5.3). Indeed, with respect to rice, there appears to have been a contradiction between the national goal of self-sufficiency[15] and reduction of poverty in the sector (Gibbons *et al.*, 1986). Overall, Lal (1975: 404) considered that despite the large volumes of expenditure directed towards the agricultural sector the results were meagre.

In 1954 a World Bank mission had advocated the development of manufacturing industry through the provision of infrastructure and tariffs. However, the Working Party on Industrial Development, appointed in 1956, took a rather different view, opting for tax incentives rather than tariff barriers. Lee (1986: 103) has noted that the Working Party's conclusion reflected the opposition to tariffs by the Agency Houses and the view that import controls would benefit the Indian and Chinese business communities and disadvantage the Malays. In the event little was done to promote industry until after the granting of full independence in 1957.

The first decisive move towards ISI was taken in 1958 with the passing of the Pioneer Industries Ordinance. This provided for tax concessions, subsidies and the setting of import and export tariffs in order to encourage industry. Initially little use was made of tariff protection (Lim, 1973: 255–57); during the early 1960s tariffs appear to have been used more to raise revenues than to promote industry (World Bank, 1963; Lee, 1986: 105–6). Following the establishment of the Tariff Advisory Board in 1963 protective tariffs were imposed at steadily increasing levels. However, until the end of the 1960s protective tariffs remained low compared to Indonesia, the Philippines and Thailand.

The Malaysian state, unlike the other economies discussed here, did not immediately become involved in production. Rather emphasis was placed on establishing a suitable environment for foreign and domestic investment. However, state involvement in production was to emerge during the 1970s in two very different ways. Firstly, in line with developments elsewhere, to pioneer production in certain areas, for example heavy industry, including cement, oil and natural gas, and to provide transport and utilities. The establishment and operation of these enterprises became an important element in state planning and direction of the industrial sector. By 1985 there were some thirty-five concerns under the jurisdiction of the Ministry of Public Enterprises.[16] Secondly, and uniquely, a series of public holding companies and partnerships were established as part of the promotion of *bumiputeras* ownership.[17]

During the early 1960s the import-substituting manufacturing sector expanded rapidly, particularly in such areas as cigarettes, soap, non-ferrous metals and motor-vehicle assembly. However, the policy was far from clearly stated and Cho (1990: 227) sees export orientation already uppermost in the planners' minds during this period. In 1966, under the First Malaysian Plan (1966–70), ISI policy was consolidated and clearly spelt out. However, by this time ISI was pressing on the limited domestic market and beginning to falter. The sector was proving incapable of absorbing the growing labour force surplus to the needs of primary production. Unemployment had increased from 2 per cent of the official labour force in 1957 to 7 per cent in 1968 and 8 per cent in 1968 (Wong, 1979: 63).

Investment grew slowly and erratically during the late 1960s, largely because of the uncertain political conditions (Saham, 1980: 26). However, foreign investment remained the principal source of funding. In 1971 80 per cent of mining, 62 per cent of manufacturing and 58 per cent of construction were foreign owned, predominantly by the UK (Hoffman and Tan, 1980: 215–16). These investments, particularly in manufacturing, formed the basis for the more rapid expansion of foreign investment and industrial production under the more favourable domestic and international conditions that prevailed during the 1970s.

In 1968 the passing of the Investment Incentive Act marked the introduction of EOI policies. These were further developed during the Second Plan (1971–5) which marked the beginning of the NEP and was geared to the establishment of an 'open', industrialised economy. It should be emphasised that EOI did not replace ISI; rather it was added to it. Production for the domestic market remained a major element in the growth of the manufacturing sector (McGee, 1986: 39). Levels of protection not only remained high but increased during the 1970s. Multi-national corporations continued to enter the ISI sector, in many cases merely replacing the import of finished goods with components and establishing highly profitable monopolies (Sundaram, 1987: 139).

During the late 1960s and early 1970s series of measures consolidated and extended the provisions for EOI. The Federal Industrial Development Agency, originally set up in 1965, was reorganised and considerably strengthened in 1969 and 1970. In 1971 provision was made for the setting up of FTZs and EPZs. By 1978 these zones employed 80,000 workers, about 11 per cent of the manufacturing labour force (*Far Eastern Economic Review* 12 May 1983: 60; see also McGee, 1986 *et al.*). In addition, in an effort to attract foreign investment, labour legislation was considerably tightened up (Sundaram, 1987: 116).[18]

Manufacturing exports began to expand steadily after 1971. The sector increased its share of export earnings from 20 per cent in 1977 to 48 per cent in 1987. However, despite the emphasis given to EOI since 1968, by 1985 less than 20 per cent of manufacturing output was exported. Of this, electronics comprised 41 per cent and textiles 14 per cent. Since the early 1970s twenty-three Japanese- and American-owned semi-conductor assembly plants have been established and Malaysia has become the largest producer in the region and the third (after Japan and the USA) in the world. Thus Malaysian industrial development has become heavily dependent on a sector that has in recent years proved footloose and far from stable and which is dominated by a small number of multi-national corporations.

In addition much of the assembly work produces very little value added for the Malaysian economy. A study of the Malaysian FTZs revealed that between 1973 and 1978, while exports totalled US$1771 million, imports cost US$1758 million, thus giving a net export surplus of only US$13 million before allowance is made for repatriation of profits, royalty payments or other fees (Lim Chin Choo, 1979, cited Sundaram, 1987: 140).[19] The FTZ developments highlight the general lack of integration between the comparatively capital-intensive export-orientated industry and the small-scale, labour-intensive domestic sector (Warr, 1986: 208).

Throughout South East Asia manufacturing has become increasingly concentrated in a small number of centres. Attempts to redress the balance have been largely ineffective and frequently half-hearted. Malaysia has, however, made more efforts in this direction than is apparent elsewhere in the region. The fear that restrictions may repel foreign investment and MNCs or reduce the competitiveness of domestically owned industry has to be balanced against the interests of *bumiputeras* and continued political stability. This was most clearly illustrated in 1975 by the adverse reaction of foreign investment to the passing of the Petroleum Development Act and the Industrial Coordination Act. They were seen as 'heralding a new and more ominous phase of Malaysian economic nationalism' (Gale, 1981: 1138). In consequence the government subsequently relaxed the controls (Hainsworth, 1979: 16).

A wide range of development agencies have been involved in the promotion of dispersed industrial activity, for example the State Economic Development Corporations and the various Regional Development Authorities. In addition, a wide range of locational incentives, the establishment of industrial estates and various regional planning measures have been used in attempts to promote a more balanced distribution of industrial activity. While Malaysian manufacturing activity has become more diffused the degree to which this has been due to the planning measures is open to question (Cho 1990). Indeed the Fifth Plan (1986–90) comments on the limited impact that dispersal policies have had appear to pave the way for the abandonment of the decentralisation policy (Malaysia, 1986: 354).

Despite weaknesses in structure and policy contradictions the manufacturing sector has become regarded by the Malaysian planners as the leading source of growth (Low, 1987: 625). This, together with the slowing of growth in the 1980s, led to two major studies being commissioned in 1983 – the Malaysian Industrial Policies Study and the Industrial Master Plan, 1986–95.

The Industrial Master Plan was published in 1986; it sets out the strategy whereby Malaysia will achieve NIC status and provides a general analysis of the structural problems of the industrial sector. Four areas of weakness were identified: excessive protection of the import substituting sector; the dominance of the export sector by electronics and textiles; the poor linkage of the export sector to the rest of the economy; and the heavy dependence on foreign capital. In addition attention was drawn to the shortage of skilled labour and technicians and, most significantly, the constraints imposed by NEP. While none of these criticisms were new it was significant that they were given official recognition. However, neither the Industrial Master Plan nor the Fifth Plan (1986–90) make any provisions for any significant changes in the industrialisation policy. Export-orientated manufacturing remains central to development but a general opening of the economy through the reduction of protection and a general overhaul of the entire industrial incentive scheme is envisaged. These developments will, however, continue to be coloured by the New Economic Policy emphasis on drastically increasing *bumiputeras* participation in industry and commerce. This may well constitute a major constraint on the proposed strategy.

Malaysian development policy since independence has had a number of distinctive features:

> the central development strategy in Malaysia must be a rare example of one promoting industrialisation and retaining a strong rural bias at the same time ... despite recognition of the need for efficiency and output growth, the development patterns have had a distinct trade-off tilted towards distributional objectives, albeit structured along a specific racial line. (Wong, 1979: 64)

These developments have rested on very high levels of government expenditure (table 5.5). During the 1970s the broad base of the Malaysian economy gave support to this; however, since 1981 the economy has encountered increasing problems. The maintenance of the high level of development expenditure in the face of these difficulties has resulted in a rapid escalation of foreign debt (table 1.18). In particular the budget deficit expanded sharply between 1979 and 1981 as the government attempted to offset the gathering recession. In 1982 the budget deficit was 3.3 per cent of GNP, one of the highest in the world (Sundaram, 1987: 125). Since 1982, and more particularly 1986, austerity measures have reduced the budget deficit (see Seawood, 1988: 52–5) but the high level of 'off-budget' expenditure, for example in connection with the public enterprises, has continued to fuel the growth of government borrowing (Sundaram, 1987: 125). Since 1986 the government has been committed to a policy of privatisation. Information on the claimed seventy-one partial disinvestments that took place during the period 1986–8 is scarce (Woon Tok Kin 1989). A large number of public concerns are loss making,[20] but their closure would be politically unacceptable and privatisation of the profitable undertakings would leave the government with little scope for cross-subsidisation (Ng Chee Yuen and Wagner, 1989: 216). The World Bank has repeatedly drawn attention to the continued high level of government expenditure and the associated escalation of foreign debt (Chandra, 1986: 44–5).

The full implementation of the NEP goal of transferring 30 per cent of the manufacturing and commercial undertaking to *bumiputeras* ownership by 1990 has become increasingly unrealistic. In addition, further moves to restrict the activities of foreign or domestic non-*bumiputeras* capital is likely to have serious consequences for further economic growth. This is particularly the case when Malaysia appears to be losing the 'competitive edge' in labour-intensive assembly work to others, notably Thailand. It may be argued that the emphasis on rural and distributional aspects of development were a necessary response to the colonial heritage. In the absence of such policies the country would have been far less stable and attracted little foreign investment. Be that as it may, the Malaysian government is now faced with serious problems of long-term economic growth and political stability. There is a basic contradiction in the construction of NEP-style development and the attraction of foreign investment. This was highlighted by the World Bank's 1986 report on the Malaysian economy, which called for an 'unshackling' of the rigid industrial-investment policy designed to enhance *bumiputera* participation (Chandra, 1986: 44).

Indonesia

Since independence Indonesian development policy has contained a strong nationalistic element, although this has varied in form and intensity. Despite major political changes there has been remarkable continuity in both the acceptance of a major economic role for the state and the importance attached to industrialisation.

Between 1950 and 1958 Indonesia attempted to disengage from the Dutch colonial economic structures through the promotion of indigenous capital (Robison, 1986: 57). By 1957 it was apparent that Indonesian capital was making little headway against Dutch and Chinese interests. In 1957 70 per cent of the plantations in Java and Sumatra were foreign owned and 19 per cent Chinese owned. Where foreign capital had been displaced it had very largely been replaced by Chinese (Robison, 1986: 42–4).

Industrialisation had been placed at the centre of the immediate post-independence development programme starting with the Economic Urgency Programme in 1951. As Wong (1979: 58) has summarised:

> Emphasis on industrial development was regarded as a rational strategy not only for bringing about fast and effective increase in per capita income but also for solving the explosive population problem in overcrowded Java.

However, little progress was made, largely due to the lack of domestic capital. Indeed, between 1953 and 1958 the manufacturing sector's share of GDP fell from 12 per cent to 11 per cent.

By 1957 Indonesia was facing a serious shortage of capital, stagnating production and declining income from exports. A retreat from economic nationalism and an opening of the economy to international and domestic Chinese capital was not a viable political option. Instead an intensified nationalist strategy was adopted with the aim of radically restructuring the economy through major state intervention.

During 1957–8 Dutch property was expropriated. This involved 90 per cent of the plantation output, 60 per cent of foreign trade, some 246 factories, and mining enterprises, as well as banks, shipping and service concerns (Anspach, 1969: 193). This property passed under direct state control, a recognition of the weakness of Indonesia's domestic capital.

Sukarno's 'Guided Economy' that emerged from 1957 was in essence state-led capitalist development. Under the policy, earnings from primary exports were to be channelled into a protected and largely state-run manufacturing sector. These funds were to be supplemented by foreign loans for the development of a heavy industrial base (Tan, 1967). The strategy was set out in the 1960 Eight Year Plan under which earnings from oil, rubber and export crops were intended to provide funds for repaying foreign loans and investment in welfare, education, public works, transport, health, food production and the manufacturing sector (Robison, 1986: 241).

Falling commodity prices and stagnant or even declining production of established staples such as rubber and sugar[21] resulted in a serious shortage of development funds and widening balance of payments deficits. Additionally, Indonesia became effectively cut off from western investment. Foreign activity only really continued in the mineral and oil sectors and investment was further restricted by the suppression of Chinese business activity.[22] Cordial relations were maintained with Japan largely because of the wartime alliance and the mediations of Sukarno's Japanese wife (Halliday and McCormack, 1973: 36). However, this relationship was not accompanied by any major influx of Japanese capital.

Buchanan (1987: 442) aptly summarised the situation:

> When Sukarno nationalised Dutch, British, US and local Chinese interests, he was asserting a double-edged anti-imperialism: firstly, the short-term advantages of using foreign investment to gain foreign exchange for development were lost, for Indonesia had neither the capital nor the expertise to develop effectively its own resources, and the institutional framework of a corrupt military, incompetent bureaucracy, and strongly capitalist entrepreneurial elite negated any possible benefits of economic independence; secondly, anti-

Western moves alienated the Western-subsidised military establishment, the trader class and the large class of commercial croppers in Sumatra.

Economic growth was extremely slow. By the early 1950s per capita income had probably reached pre-war levels but overall growth for the period 1955–65 was below that of population (Glassburner, 1971). Per capita income fell from US$73 in 1960 to US$65 in 1963. In 1960 79 per cent of export earnings came from rubber (47 per cent), oil (27 per cent) and tin (4 per cent). Of these only oil was showing significant growth. The manufacturing sector, despite the emphasis placed on it, made by far the smallest contribution to the economy of any of the future members of ASEAN. As in Thailand, most accounts blame the poor performance of the state-run industrial sector on the efficiency of management, corruption and misuse of funds (Fryer, 1979: 321). While on the whole state enterprises were failures, some did operate efficiently and profitably (Robison, 1986: 74). Insufficient attention has focused on the difficult internal and international situation that Indonesia faced. However, by the early 1960s the economy was in a state of virtual collapse (Fisher, 1971: 303). Per capita income and food intake were amongst the lowest in the world. Per capita rice production had fallen below the already inadequate levels of the 1930s (Hardjona, 1983: 47). The balance of payments was deteriorating rapidly, overseas debt rising steeply and, by 1965, inflation running at 594 per cent (Palmer, 1978: 6–23).

The sheer size, physical fragmentation and cultural heterogeneity of Indonesia, reinforced by uneven development during the colonial period, posed serious problems for the newly independent state. For most of the period 1957–65 Sukarno managed successfully to play off the two main political power blocks, the PKI (Partai Komunis Indonesia) and the military. However, development was increasingly hindered by political and social unrest, of which the revolutionary and successionist movement in the outer islands were only one manifestation. Despite increased military expenditure and involvement in provincial administration the outer islands, particularly Sumatra, began to trade directly with Malaysia and Singapore, thus depriving the central government of revenue.

The Sukarno government faced serious economic and political problems and the regime's failings were many. However, the Indonesia case, like that of Burma, highlights the difficulty that a former colonial country faces when it attempts to develop largely in isolation from the world economy. Much criticism has been directed at the regimes of both these countries for their failure to harness the resources to generate rapid economic growth (see for example Fryer, 1979: 23–34). It is perhaps the whole question of disengagement from the world-economy that should concern us more, particularly in view of the almost certain Western involvement in the overthrow of the Sukarno government.[23]

The bloody end of the Sukarno regime in 1965–7 and the establishment of a pro-Western, military government under General Suharto heralded a period of rapid integration into the world-economy. A key element in the new regime was a group of US-trained 'technocrats' who became known as the 'Berkeley Mafia' (Ransome, 1970). Centring on this group, the 'New Order' adopted a policy of maximising economic growth, foreign investment and industrial development.

The importance to the interests of international capital of Indonesia's economic rehabilitation and full reincorporation into the world-economy may be judged by the reaction of the Western nations and the international agencies. In 1966, shortly after Suharto came to power, what can only be described as an international economic rescue operation was mounted. This involved debt rescheduling and a massive amount of aid.

Negotiations between the Western creditors, principal aid donors, the IMF and IBRD resulted in the formation of the IGGI (Inter-Governmental Group on Indonesia). This was to coordinate and channel Indonesian loans on increasingly easy terms. The volume of aid grew rapidly from US$0.64 billion in 1971–2, to US$0.88 billion in 1973–4 and US$1.3 billion in 1976–7. In 1968 the IBRD took the unprecedented step of setting up a permanent mission in Jakarta and began to work very closely with the national planning apparatus (Palmer, 1978: 31–3). The subsequent national plans were not only very heavily funded by loans channelled through the IGGI but donors had a very direct impact on the planning process.[24] Thus Indonesia was effectively stabilised and the conditions established for the operation of international capital.

As Wong (1979: 59) has noted, the problems facing the new regime were similar to those facing Sukarno in the early 1950s: rapid population growth, growing un- and under-employment and low per capita income.[25] Rapid industrialisation was again presented as the solution, and became a key element in subsequent national plans. The New Foreign Investment Law (PMA) of 1967 and New Domestic Investment Law (PMDA) of 1968 heralded a period of rapid growth in industrial production and foreign investment.[26] By 1970 the Chairman of the Indonesian Government Investment Committee was able to state that: 'Japan and the US already control the Indonesian Economy. The US has seized the natural resources and Japan the manufacturing industry' (cited Halliday and McCormack, 1973: 38). While there is an element of exaggeration in this, for it took several years to establish a suitable climate for foreign investment, it does reflect a very clear trend in the early 1970s. Between 1967 and 1985 58 per cent of investment in the oil sector was American and 65 per cent of the manufacturing investment Japanese (Bank of Indonesia and American Embassy data cited by Hill, 1988: 55 and 59).

Since 1968 the exploitation of oil has dominated foreign investment in Indonesia.[27] Petroleum's share of foreign investment was 59.1 per cent in 1968, falling to 40.8 per cent in 1972, as investment moved into other primary sectors and manufacturing. From 1973 petroleum's share increased rapidly, particularly following the 1974 and 1979 oil price rises, reaching 70.0 per cent in 1979 and 85.5 per cent in 1982.

The first wave of non-oil investment that followed the passing of the PMA in 1967 was attracted by Indonesia's resource base. In 1970 40.1 per cent of investment went into minerals and 29.1 per cent into forestry, principally in the outer islands. The speed of foreign penetration was remarkable. In the forestry sector, for example, between 1967 and 1969 twenty concerns invested US$165 million and received 5.8 million hectares of concessions – 5 per cent of the nation's reserves (Palmer, 1978: 122).[28] However, domestic capital was attracted by the high rates of return and the sheer size of the reserves (120 million hectares) meant that the concessions could not easily be cornered by foreign concerns. As a result by 1973 42 per cent of the forestry investment was domestic. However, foreign concerns came rapidly to dominate mining, supplying 92 per cent of investment by 1973. In addition, between 1971 and 1976 foreign companies began to return to the estate sector, either in partnership deals with the government or by complete repossession in return for meeting the full rehabilitation costs (Palmer, 1978: 119).

Foreign investment in the primary production was of major importance in the sector's rehabilitation and expansion. In many areas growth was spectacular. Timber exports, for example, increased from 1 per cent of export earnings in the early 1960s to 10 per cent in 1974. However, this rapid removal of Indonesia's resources by MNCs to supply the needs of industry elsewhere pose serious long-term problems for the economy. From the early 1970s, while foreign investment in primary production continued to expand, the sector's share fell, as manufacturing increased its share of non-oil investment from 22.5 per cent in 1970 to 73.0 per cent in 1976 (Bank of Indonesia, 1982).[29] This switch towards investment in manufacturing was accompanied by increased emphasis on the Inner Islands, particularly the Jakarta area.

During the colonial period Indonesia's limited manufacturing sector was heavily concentrated in Java. Foreign investment in manufacturing has reinforced this uneven pattern (Hill, 1988: 40–5); between 1967 and 1979, 73 per cent went to the West Java and Jakarta core (Forbes, 1986: 123–7). In 1985 Java had 86 per cent of manufacturing employment and 78 per cent of large- and medium-sized firms. Within Java there is a pronounced concentration of large- and medium-sized firms in West Java and Jakarta (Forbes, 1986: 127). Lesser concentrations are found in Central Java (weaving, batik and tobacco), East Java (weaving, tobacco and cigarettes), East Kalimantan (timber) and North Sumatra (rubber and timber).

The rapid integration of Indonesia into the world-economy after 1967 was a more complex and antagonist process than the account so far would suggest. Unlike the Philippines, where indigenous capitalist and foreign investors were both generally well served by the accommodation that was made to international capital, in particular during the Marcos period, in Indonesia their interests have frequently conflicted.

From 1968 until 1974 the Suharto regime was beholden to the US-backed international agencies in particular, and foreign investors in general for 'resourcing' the economy. During this period, largely because of economic weakness, there was a retreat from economic nationalism and state participation. The economy was reconstructed around foreign investment and loans. Foreign investment share of total capital formation rose from 11.7 per cent in 1968 to 31.8 per cent in 1971 (Hill, 1988: 35).

The sharp rise in the price of oil in 1973–4 gave Indonesia the financial basis to reestablish state-led development. By the early 1980s oil and gas sales represented over 80 per cent of export earnings and 70 per cent of government revenue. The economy enjoyed substantial balance of payment surpluses (figure 1.4).

Initially the relationship between development and oil revenues was very direct with Pertamina (the state oil corporation) functioning as a para-statal organisation with independent ability to raise foreign loans and a budget which peaked at 50 per cent of GDP. A number of the large industrial undertakings that came into operation during the mid-1970s were Pertamina projects – for example, the PURI fertilizer plant at Palemlang. Indeed during Repelita I (1969–74) it is scarcely an exaggeration to suggest that there were two coordinated national development programmes in operation (Bartlett *et al*., 1972). However, the collapse of Pertamina in the mid-1970s revealed corruption and misuse of funds on a massive scale (McCawley, 1978).

By the mid-1970s the availability of oil revenues and cheap international loans had established the dominance of state development. As Robison (1987: 5) has summarised:

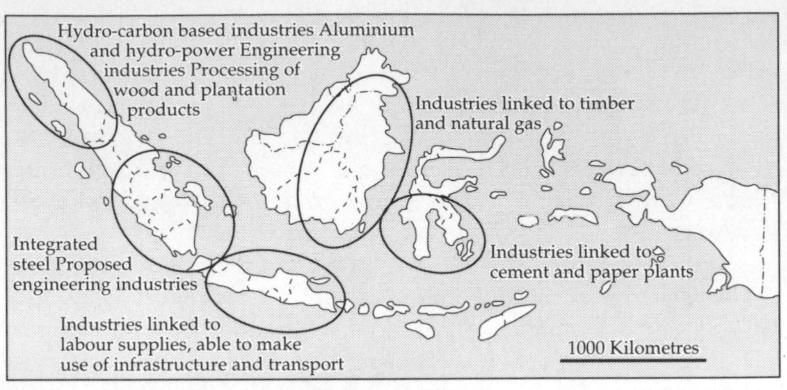

Map 5.1 Indonesian special development areas
Source: Forbes, 1986: 137

The expanding domestic corporate conglomerates, closely tied in the centres of military bureaucratic power, including the presidential palace and the Murtopo group, were building their empires upon access to contracts, credits and monopolies made possible by state investment fuelled by oil income. Policies of state-led import-substitution industrialisation had now become integral to the interests of the alliance of economic and politico-bureaucratic power.

To protect the interests of domestic capital foreign investment was, from the mid-1970s increasingly restricted.[30] For example, state bank credit was limited to indigenous companies; within ten years companies investing under PMDN and PMA had respectively to be 75 per cent and 51 per cent in indigenous hands. Effectively many sectors of the economy became closed to foreign investment and were reserved wholly or partially, for domestic capital (Palmer, 1978: 102–5). Despite these restrictions, some close relationships between domestic industrial groups and international capital continued to develop. However, increasingly the nationalist policies were running contrary to the interests of international capitalism and foreign investment fell from the peak of 1975 (figure 1.1).

As in Malaysia and Thailand, the Indonesian state has attempted to redress the uneven pattern of development. However, there was no coordinated policy until the mid-1970s (Donges, Stecher and Wolter, 1980: 363). Since then three measures, the BKPM Investment Priorities list, the establishment of industrial zones (map 5.1) and the development of EPZs, have been used in an attempt to develop industry in rural areas and outside of Java. To date these programmes have been of limited success. The outer islands remain attractive only to concerns linked directly to primary production.

Table 5.10 *Indonesian development budget, 1974–89, percentage share of sectors*

	Repelita II[1] (1974–79)	Repelita III[2] (1979–84)	Repelita IV[3] (1984–89)
Agriculture and irrigation	19.1	14.0	12.9
Industry	3.5	5.4	5.4
Mining and energy	7.4	13.5	15.3
Transport and tourism	15.8	15.5	12.3
Trade and co-operatives	0.7	0.9	1.2
Manpower and transmigration	1.3	5.7	5.9
Regional development	17.7	9.8	6.9
Religion }	10.3	0.7	0.7
Education and youth affairs }		10.4	14.7
Health and family planning	3.7	3.8	4.4
Housing	1.9	2.4	3.8
Law	0.0	0.9	0.8
Defence and security	3.0	6.8	6.7
Information	0.5	0.7	0.6
Science and technology	1.9	2.0	2.3
Government apparatus	2.3	2.7	1.3
Business development	10.7[4]	1.7	2.2
Environment	0.0	3.2	2.5

[1] US $12,650 m. at 1973 exchange rate (US $1 = 415 rupiahs)
[2] US $34,960 m. at 1979 exchange rate (US $1 = 625 rupiahs)
[3] US $77,741,000 m. at current prices
[4] Government capital participation
Source: *Far East and Australia*, 1987: 443

The mass poverty of rural Java (see section 4.4) has been a focus of government policy since independence. However, rural development has become increasingly orientated towards the raising of agricultural productivity in order to reduce the level of rice imports. The various policies implemented during the Sukarno period met with little success (Hardjono, 1983: 47–8). Renewed and more sustained efforts since 1967 have proved far more effective. These have involved the rehabilitation and extension of the irrigation systems constructed during the Dutch period and the institutionalisation of the BIMAS (Mass Guidance) and INMAS (Mass Intensification) programmes which have made subsidised credit, fertiliser, insecticide and extension services widely available. Principally as a result of these programmes and the commercialisation of rice production which accompanied their implementation, paddy output rose from 11.7m tons in 1968 to 26.3m tons in 1986. Rice imports fell from 1.9m tons in 1979 to *c.* 30,000 tons in 1986. However, the intensification has, in many areas, seriously disrupted rural society, increased landlessness and unemployment, reduced neither the depth or incidence of poverty (see section 5.3) and fuelled the already high levels of rural-urban migration (Collier, 1975; 1978; Hardjono, 1983; Gibbons *et al.* 1986).

The main thrust of Indonesian policy towards the extreme imbalance between Java and the outer islands has centred on 'transmigration' and regional development programmes. As may be seen from table 5.10 regional policies' share of the development budget has fallen steadily since Repelita II while transmigration has been given increased prominence.

Resettlement of people in the outer islands as a means of reducing population pressure in Java, Bali and Lambok was by no means a new policy. In essence the programmes implemented between 1950 and 1970 represented a resumption of the Dutch colonisation programme (chapter 4).

The programmes have rested on the establishment of villages based on irrigated rice growing. In general the conditions were unsuitable for such development, particularly in the absence of funds for large-scale irrigation development (Hardjono, 1983: 55). Unrealistically high targets (table 5.11) and inadequate planning undermined the credibility of the whole programme.

Under Repelita II (1976–9) and, more especially, Repelita III (1979–84) a revised transmigration scheme was given prominence. Procedures have been streamlined and support services expanded. These developments owe much to World Bank support and funding. Resettlement has been increasingly orientated towards the cultivation of non-rice cash crops. However, as may be seen from table 5.11, the total number of people resettled has been insufficient to make any significant impact on the population of the inner islands, and much of the outer island development has proved less than sustainable. Indeed increasing concern is being raised over long-term environmental damage in the outer islands (Hardjona, 1983: 57). The latter issue was also raised by the World Bank (1988) in a report highly critical of the whole transmigration programme. This study recommended that future development should concentrate on the consolidation of existing schemes rather than further sponsored migration. The acceptance of these proposals by the Indonesian government resulted in a US$250 million loan for 'second stage development' of the schemes (Vatikiotis, 1987: 28; World Bank, 1988b).

Between 1967 and 1975 the manufacturing sector grew at an average annual rate of 16.5 per cent. The subsequent slowing reflected the saturation of the domestic market and the slow move into intermediate and capital goods. Consumer goods fell from 80.8 per cent of manufac-

Table 5.11(a) *Indonesia, sponsored transmigration programme, 1950–84*

Year	Total families moved	Local families[1]	Resettled families[2]	Total families settled	Total people
1950–54	21,037	0	1,280	22,317	87,000
1955–59	32,114	0	128	32,242	134,000
1960–64	26,456	0	0	26,456	111,000
1965–69	21,633	0	0	21,633	92,000
1969–74 (Repelita I)	39,436	0	75	39,511	240,000
1974–79 (Repelita II)	44,484	7,600	0	52,084	465,000
1979–84 (Repelita III)	301,279	22,284	42,414	365,977[3]	1,492,000
TOTAL	486,439	29,974	43,897	560,220	2,320,000

[1] Indigenous families settled in transmigration sites
[2] Resettlement of sponsored or spontaneous migrants within the province
[3] In addition, 170,000 families were identified that moved spontaneously

Table 5.11(b) *Destination of sponsored migrants, 1950–84*

	Sumatra	Kalimantan	Sulawesi[1]	Irian Jaya[2]	Total
1950–54	20,400	1,400	500		22,300
1955–59	28,900	2,600	700		32,200
1960–64	21,000	4,500	1,000		26,500
1965–69	16,500	2,100	2,700	300	21,600
1970–74	22,000	6,000	11,400	100	39,500
1975–79	33,000	11,000	9,000	2,000	55,000
1980–84	227,100	70,600	51,700	16,600	366,000
Total	368,900	99,200	77,000	19,000	563,100
Percentage	65	18	14	3	100

[1] Includes the Moluccas (Malukus) and other small eastern islands
[2] Not incorporated into Indonesia until 1963
Source: World Bank (1988): XXI and XXIII

turing output value in 1971 to 47.6 per cent in 1980 (Robison, 1987: 26). Despite these developments the percentage contribution of manufacturing to national income and exports remains the smallest of the ASEAN group (Hill, 1988: 13–27; table 5, 1.12 and 1.14).

Principally Indonesia was able to finance industrial development and adopt, in some instances, a cavalier attitude to MNCs and international capital interests because of the level of oil revenues. However, with the decline in oil prices during the 1980s the whole system has come under increasing pressure.

Since 1981 Indonesia has been faced by balance of payments deficits which have been met by rapid increases in overseas borrowing (figure 1.4 and table 1.18). Contraction of domestic and overseas markets has resulted in large numbers of plant closures and marked over capacity in a wide range of industries (Far Eastern Economic Review 20th January 1983: 47–8).

The World Bank (1981, 1984 and 1985) has become increasingly critical of Indonesian ISI, advocating, as it has done for most Third World countries, an opening up of the economy to international capital and the removal of 'distortions to the market' in the form of subsidies.[31] Particularly criticism has been directed at the high-cost nature of much Indonesian industry. While labour costs are perhaps the lowest in the region (tables 1.2 and 6.1) the goods produced are frequently too expensive to compete in export markets.

The Indonesian government has responded to the crisis and the pressure from the international agencies by reforming tax and excise regimes, reducing subsidies on fuel and basic foodstuffs, axing a number of major industrial projects, removing credit and interest-rate ceilings to encourage domestic capital accumulation and various export incentives were introduced. However, these measures do not affect the centrality of the state-led ISI.[32] The continued ability of the Indonesian state to resist the pressures to dismantle the 'economic nationalism' structure ultimately depends on the level of export earnings. In the absence of a major recovery in the price of oil these are increasingly derived from rubber, tin, timber and palm oil. During 1987–8 demand for these products was comparatively buoyant in the wake of a limited recovery amongst the developed economies. However, renewed dependence on these non-oil primary exports will increase the vulnerability of the Indonesian economy, tying it much more closely to the fortunes of the major raw material importers.

The Philippines

The development of the Philippine economy since independence has been coloured by the country's continued dominance by the USA, recurrent economic crises and, from 1961, subservience to the IMF. Nowhere else in South East Asia has the influence of the former colonial power remained so strong.[33]

The wartime destruction of the economy forced the Philippines into a dependence on foreign capital for reconstruction:

> The United States took advantage of the devastated condition of the Philippines to blackmail the newly independent republic into surrendering the reins of its external trade and monetary policies in exchange for much needed aid. She then re-gained the pre-war preferences and secured unrestricted access for American capital to the country's raw materials and domestic markets. (Canlas et al., 1988: 20)

In 1946 American corporations and individuals were given the same rights as Filipinos with respect to ownership and exploitation of natural resources. Domestic capital was given no preferential treatment and indeed was put in an increasingly disadvantaged position. Under the prevailing free-trade situation domestic manufacturing was inhibited by the influx of cheap American produce (Wong, 1979: 68). Essentially between 1946 and 1949 the pre-war economic relationship with the USA was substantially reimposed.

In 1949, following a major foreign-exchange crisis the USA allowed the Philippine government to establish exchange and import controls (Canlas et al., 1988: 21). These were later reinforced by a high level of selective import tariffs to provide the framework for ISI.

Between 1950 and 1962 the manufacturing sector, starting from a very low base, grew at an average rate of 10.6 per cent compared to 6.2 per cent for the economy as a whole. The sector's share of GDP increased from 8 per cent in 1949 to 19.3 in 1960 and of employment from 6.6 per cent in 1952 to 12.1 per cent in 1961 (World Bank, 1976: 527 and 529; Valdepenãs, 1970: 141). The main growth was in assembly, packaging and processing concerns. These were heavily concentrated in the Manila area because of the size of the market and their dependence on imports (World Bank, 1976: 187). Much of this development was financed by American capital which, as Golay (1961: 107) has noted, was attracted by the 'highly protective and lucrative Philippine market'. However, American investment was to a degree inhibited by the exchange controls which limited the repatriation of profits and capital (Canlas et al., 1988: 23).

Despite the development of consumer goods protection during the 1950s the manufacturing sector remained dominated by the simple processing of primary produce, particularly sugar, wood, coconut oil, gold and silver. In 1960 primary products, mainly timber, copra, bananas, abaca and tobacco, accounted for 49.3 per cent of export earnings, while processed primary goods accounted for a further 48.4 per cent (World Bank, 1976: 221).

Growth of the manufacturing sector levelled off from 1960 with the saturation of the limited domestic market. In addition deteriorating terms of trade perpetuated a serious balance of payments crisis. In 1961–2 the USA and the international agencies took a very hard line, forcing devaluation, the removal of exchange controls, liberalisation of import policies and arranging a US$300 million IMF loan (Stauffer, 1985: 248). While these policies initially resulted in virtual stagnation of the economy and soaring consumer prices they also opened the way to much fuller involvement of foreign, particularly American, capital. In the wake of the 1962 devaluation many bankrupt or ailing Filipino-owned concerns were purchased by American based MNCs.

The rapid increase in the number of American-owned branch plants during the early 1960s was accompanied by the development of export markets. The share of manufactured goods in export earnings increased from 4 per cent in 1960 to 12 per cent in 1965. This expansion was greatly accelerated after Ferdinand Marcos came to power in 1965. During the late 1960s Marcos led a campaign to entice more American and Japanese investment (Lim, 1985: 249). In 1967 the Investment Incentive Act was passed and the Board of Investment established both with the aim of encouraging foreign investment and EOI. The incentives for export-oriented industry were under the 1970 Export Incentive Act. In 1970, on the advice of IBRD, the first EPZ in Asia was established on the outskirts of Manila. Four more were established by the early 1980s, by which time they employed 30,574 workers.

In addition, under the repressive Marcos regime, strikes were made illegal in 'essential industries', and virtually so elsewhere. The establishment of the corporatist Trade Union Congress imposed compulsory arbitration and seriously restricted wage levels, workers' rights and conditions of employment. Between 1972 and 1980 wage levels fell by 42.6 per cent for unskilled workers and 32.8 for skilled (Philippines, National Economic Development Authority, 1980: 14). Thus MNCs were presented with a cheap 'trouble free' labour force.

From the early 1960s increasing amounts of foreign investment moved not only into manufacturing but also into primary production and associated processing. There has been rapid expansion of MNC involvement in fisheries, oil palm, rubber, pineapples and bananas, complementing the long-established presence in sugar and copra. While the main source of foreign investment was American there have been limited and more recent incursions of Japanese capital. As with the manufacturing industry, many of the agri-business MNCs arranged partnership deals with local capital and the state. However, even where subsidiaries were set up using large proportions of local capital, effective control remained vested in the parent MNC. The process of MNC agri-business development in the Philippines is very well illustrated by Krinks's (1986) study of the banana industry.

During the 1960s Castle and Cooke, United Fruit and Del Monte established a dominant position in the Philippines' banana industry. Initially these developments centred on the locational advantages of the islands for penetrating the Japanese market.[34] The domination of the industry by the major American based MNCs was often established through a complex set of subsidiaries and purchasing contracts.[35]

MNC agri-business has, particularly since the early 1960s, been encouraged to acquire very large estates. A wide variety of incentives and concessions have been made available to encourage large-scale cultivation and stock rearing, particularly when orientated towards the export market. In many instances land has been leased from small cultivators. However, 'those who were unwilling to lease came under heavy pressure, including threats of violence, and the banana industry in particular stimulated considerable unrest that has continued' (Krinks, 1983: 112). In total many thousands of hectares have passed from peasant cultivators into the hands of MNCs.

Persistent unrest has focused the attention of successive governments on the rural areas. Insurgency has been a continual threat to the extraction of surplus from the rural areas and, if not to the continuation of the prevailing system, certainly to the inflow of foreign investment. Since 1950 a series of rural development programmes have been undertaken, initially heavily reliant on American aid. During the 1950s much expenditure was directed at counter-insurgency and strategic road building (Constantino and Constantino, 1978: 230).

Successive governments and the international agencies have agreed that rural unrest is a direct consequence of poverty resulting from the land-holding system (World Bank, 1976: 475). Since 1954 a series of land-reform programmes and measures have been initiated. To date their impact has been limited. Indeed the 'progress' in land reform has to be balanced against the loss of land by small farmers to the interests of the multinational agri-business. The considerable political power of the large land owners has thwarted, watered-down or delayed every measure that has been introduced. By 1986, for example, only 13,590 tenant farmers had received titles to 11,087 hectares of land; it was estimated that 80 per cent of the land was owned by 20 per cent of rural households (Philippines, National Economic Development Authority, 1986). There is no reason to think that the present Aquino regime will achieve significantly more in this respect.

In the Philippines, as in Malaysia and Indonesia, attention has been focused on the rural sector because of the need to reduce the dependence on food imports. By 1965 the stock of uncultivated land was almost exhausted, agricultural productivity was static and food imports represented 19 per cent of the import bill (World Bank, 1976: 101–2 and 540).

Rural development funds became heavily concentrated on raising agricultural productivity. The recipients of these programmes were principally large landholders. This was particularly the case in the rice- and sugar-growing areas (World Bank, 1976: 161). In addition items such as subsidised fertiliser, earmarked for the domestic food sector, were directed to export crops. However, the expansion of production was impressive, a precarious self-sufficiency in rice and maize being achieved by 1978.

The price of rice has, since 1970, been kept artificially low, thus keeping down the cost of living of the urban population (and thereby wage levels) and depressing rural income levels. The rural sector has very largely been viewed as a source of food, raw materials and export earnings to finance urban-industrial and infra-structural developments. Over the last thirty-five years successive governments have laid emphasis on 'social justice' and reduction of rural poverty (Krinks, 1983: 101). These statements are belied by the actions of the government and the lack of evidence for any major reduction in poverty or increased equality (see section 5.3). While these attitudes and policies are found elsewhere in the region they have been at their most extreme in the Philippines.

The Marcos regime was far from the open, free-market image that has been projected. During the 1970s the state became involved in virtually every part of the economy. State participation in industrial development was widespread with, apart from the complex provision of incentives for MNCs and construction of infra-structure, increasing direct involvement in production. A complex and highly corrupt alliance of state, domestic capital and MNCs emerged. This 'crony capitalism' as many termed it (Bonner, 1987), reflected the basic weakness of the Filipino elite. However, during the 1970s there is little doubt that both international and domestic capital derived considerable benefits from these associations.

From the mid-1970s the Philippines, like other oil-importing primary producers, became increasingly dependent on foreign loans. This was reinforced by the need of the Marcos regime to maintain high levels of expenditure in order to preserve its power base. Austerity measures would have been politically dangerous. In addition, the government, with the active support of American advisers and the international agencies, opted for a 'Brazilian-style' debt-led growth strategy (Jayasuriya, 1987: 93). In consequence the budget deficit widened from 0.6 per cent of GNP in the early 1970s to 4.3 per cent in 1983. Thus foreign debt and the debt-service ratio were already rising sharply before the 1979 oil price rise and the crises of the 1980s. As international credit became increasingly scarce and expensive the instability of the Filipino economic and political structures became exposed. By 1981 the Marcos regime was coming into increasing conflict with the IMF and IBRD over the implementation of structural reforms.

From the early 1970s manufacturing's share of export earnings increased rapidly, rising from 6 per cent in 1965 to 50 per cent in 1980. However, the sector's share of GDP and employment showed little expansion, indicating the limited impact that EOI has had on the domestic economy. The ephemeral nature of much of the EOI development is revealed in the rapid withdrawal of foreign investment during the economic and political crises that affected the Philippines during the 1980s. Similarly there have been violent fluctuations in the value of manufactured exports, in particular of semi-conductors which increased by 35 per cent in 1983 and fell by 43 per cent in 1984. The development of EOI during the 1970s did not result in any fall in the level of protection for domestic industry. Despite pressure from the World Bank to dismantle the protectionist structure, no real action took place. Indeed by the mid-1980s the Philippines had the highest level of tariff protection in ASEAN – over 70 per cent, still at the level of the mid-1970s (Jayasuriya, 1987: 85). The continuation of these levels of protection served the interests of international as well as domestic capital because of the heavy involvement of American branch plants in the domestic market (Jayasuriya, 1987: 95).[36]

Only in 1980 did the government agree to reduce tariff protection as part of the conditions of a US$600 million IBRD Structural Adjustment Loan. Plans were laid to reduce the average level of protection from 42 per cent to 28 per cent. The Marcos regime experienced considerable difficulty in keeping to the agreed timetable of tariff reform due to the vested interests involved – including those of US-owned branch plants. When the regime collapsed in 1983 virtually no progress had been made (Jayasuriya, 1987: 85). The Aquino administration has encountered similar difficulties. The government has come under increasing pressure from IBRD and the IMF to phase out protection. However, the pressures of vested interest remain and while, during 1986 and 1987, quantitative controls (quotas, licensing or bans) were removed on 1232 products (leaving 303 protected items)[37] they have been replaced by tariff covers of up to 50 per cent (*Far Eastern Economic Review Year Book*, 1988: 232).

Despite the incentives and the privileged position that foreign capital occupied in the Philippines the level of foreign investment remained well below the levels of Malaysia, Indonesia and, of course, Singapore. However, investment figures almost certainly underestimated the degree to which manufacturing investment was undertaken by overseas' interests. Foreign firms appear to have raised the bulk of their capital in the Philippine capital market (Lindsey, 1983). In the Btaan EPZ, Warr (1984) found that 90 per cent of the capital invested was raised locally. There is little doubt that the reluctance of foreign investors to invest in the Philippines reflected their appraisal of the country's long-term stability. However, it is also the case that labour productivity in the manufacturing sector was low by regional standards (World Bank, 1976: 196) and the wage levels higher than Indonesia and Thailand (tables 1.2 and 6.1). Poor infrastructure, notably inadequate and unreliable power generation, have further discouraged investors (Galang, 1988). In addition, many investors were discouraged by the complex and ever-changing bureaucratic regulations governing foreign investment and the difficulty of operating the corrupt atmosphere of the 'crony capitalism' that developed during the Marcos period.

During the 1980s the international recession and increasing political instability brought the economy of the Philippines to the verge of collapse. Since the Aquino government came to power in 1986 a measure of economic stability has been achieved through debt rescheduling and the extension of IMF and World Bank credit facilities. There have been limited signs of renewed confidence on the part of foreign investors and slow, but uncertain, economic growth. Per capita GNP, which fell by 20 per cent between 1978–9 and 1985–6, had recovered to 90 per cent of the 1979–80 level by early 1989. The political situation remains extremely uncertain and the limited economic recovery has to be balanced against deteriorating balance of payments and increasing foreign debt.

5.3 South East Asian development: an evaluation

The changes in the South East Asian economies over the last thirty years have been spectacular (table 1.12 and 1.14). Substantial industrial sectors have been established and in all except Indonesia the dependence on primary exports has been reduced. It is questionable whether these changes can be attributed to the development strategies *per se*. Rather the policies have to be seen in terms of *facilitating* changes made possible by the development of the international economy.

While the general form of development shows great similarities across the region there have been differences in timing and emphasis. In addition, the disparities between the region's economies have tended to increase (chapter 1). However, there is little to be gained by reviewing the relative 'successes' of the individual states. A ranking which puts Singapore at the top and perhaps the Philippines at the bottom tells us little – certainly not that the Philippines adopted the 'wrong' strategy and Singapore the 'right' one.

In general, development strategies have become increasingly dependent on foreign investment. However, the growth of indigenous capitalist classes has also been fostered, and while interests of domestic and international capital frequently coincide they have come increasingly to conflict. The resolution of these conflicts depends on the relative strength of domestic capital.

In response to their heritage of uneven development all the countries in the region have adopted spatial planning strategies. While the Malaysian efforts and achievements have perhaps been the most impressive, nowhere in the region has development become markedly more even.[38] In consequence much criticism has been levelled at spatial planning mechanisms, structures and policies; less attention has focused on the context in which these operate. At the simplest level, ISI and EOI, whether domestically or foreign financed, have resulted in rapid growth in already developed locations, the Manila area, Jakarta and West Java, and Bangkok – Thonburi. These areas have the greatest concentration of buying power and most suitable labour force infra-structure, including port facilities. The more internationalised the economies have become the more foreign investment and MNC activity emphasise the unevenness of development. Thus attempts to redress spatial imbalance generally run counter to the process of development engendered by domestic and international capital.

As we have seen in Malaysia and Indonesia attempts to direct the location of economic activity frequently discourage foreign investment. In general, the use of incentives, including industrial estates and FTZs, to attract industry to backward areas has met with little success. The failure of these policies has been used as a major argument for their dismantling. Under the economic conditions facing the South East Asian nations during the 1980s there has been a general retreat from spatial planning, most recently in Malaysia.

A more diffused pattern of growth has resulted from developments based on natural resource exploitation. However, while incentives have been utilised to attract capital into these sectors there has been little attempt to control its location. Indeed, to state the obvious, this has reflected the distribution of accessible resources. Much of this development has been very short-lived, as with forestry in the Philippines and Indonesia. More long-term mineral and agricultural developments have tended to be capital-intensive, generating little local employment or linkages with the local economy. A number of writers have commented on the speed with which resources are being removed from South East

Asia by MNC with little long-term benefit occurring to the local economies (Canlas *et al.*, 1988: 24). The depletion of natural resources is now seen as a serious long-term threat to export earnings and industrial development.

Despite the development of the industrial and service sectors the majority of South East Asia's population remain directly dependent on primary production. Urban-industrial development remains underpinned by food and raw material production. As well as low and uncertain price levels the primary export sector is now facing the exhaustion of many key resources and rising domestic consumption. While this is most clear with respect to oil, forestry and fishery products there is now widespread evidence of land shortages, environmental degradation and faltering agricultural production.

It is estimated that Indonesian oil production will level off at *c.* 700 million barrels (Robison, 1987: 377). Thus by the end of the present century there would be only 200 million barrels available for export compared to a peak of 470 million barrels in 1978. The sustainability of even this level of exports is open to question. Current estimates suggest that reserves will begin to run out early in the twenty-first century (Mackrell, 1986).

South East Asia remains a major exporter of timber and wood products. However, serious national, regional and local shortages have now emerged. Most strikingly Thailand has, since 1978, been a net importer of wood. Similarly Java has become heavily dependent on supplies of timber and fuel wood from the outer islands. The region's remaining forestry reserves are heavily concentrated in Indonesia (55 per cent), predominantly Irian Jaya, Burma (15 per cent) and Malaysia (10 per cent), especially Sabah and Sarawak. Given the rate of deforestation (350,000 hectares per annum during the 1980s), the general ineffectiveness of conservation measures and the limited replanting (less than 16 per cent of cutting between 1980 and 1985), the long-term prospects for timber exports and the timber-processing industry are far from bright (Aditjondin, 1984; Rao, 1984).

Similar arguments can be advanced for the fishery sector where stock depletion has been greatly accelerated by the growth of canned and frozen seafood exports and increasing coastal pollution (Kurien, 1984).

The rapid expansion of the urban industrial sector has produced major problems of pollution, congestion and environmental damage. These problems are at their most extreme in Jakarta and Bangkok. In the latter, excessive extraction of underground water has compounded these problems by causing widespread subsidence and wet-season flooding (ESCAP, 1988). To a degree the rapid growth of the manufacturing sector and the lack of effective controls has resulted in the importation of the environmental consequences of industrialisation from the developed economies.

Environmental and resource conservation issues are becoming increasingly matters of debate in South East Asia. However, while all the governments have adopted a range of 'environmental policies', their impact has been slight. Many of the programmes, for example the Thai 'Greening of Isan' (reafforestation of the North East), are likely to founder on the scale of the task and a lack of funds.

All the states have over the last thirty years reiterated their commitment to the eradication of poverty, particularly in rural areas. However, this has generally run counter to the emphasis on urban-industrial development that has dominated development strategies. High incidences of poverty persist in the region (table 5.12). As with regional disparities there is little evidence that points conclusively towards a general reduction in the depth or incidence of poverty.

The evidence for increases in the incidence of poverty appears to be most conclusive for the Philippines. A range of studies point to an increasing proportion of the population falling below the variously drawn poverty lines.[39] The World Bank (1980 and 1988b) reported that the percentage of families living in poverty rose from 43.3 in 1964 to 44 in 1975 and to 52 in 1985. In addition the 1988 study concluded that real wages had dropped constantly since 1960. Canlas *et al*. (1988: 13) cites a study by a private research organisation (IBON Database Philippines 1983) which suggested that 71 per cent of all families were living in poverty.[40] Further, a number of studies point to increasing income inequality, with, for example, an increase in the Gini coefficient[41] from 0.51 in 1965 (Montes, 1985: 634) to 0.60 in 1975 (Mukhopadhyay, 1985: 10). Similarly between the 1971 National Census and Statistical Office survey of family income and the 1979 Integrated Survey for Households the income share of the poorest 60 per cent of households declined from 25 per cent to 22.5 per cent.

In Thailand a more complex situation is revealed by the National Statistical Office Household Expenditure Surveys (table 5.13). While the incidence of poverty appears to have generally declined at the national level and in rural areas, its incidence has increased in provincial urban centres. Income disparities, measured by Gini coefficients, have increased except in the urban centres of the South and North East. In terms of regional incomes, despite the importance attached to this issue in the first three national plans, there is no evidence of convergence. Indeed, official statistics (table 5.14) suggest increasing disparities, particularly between the North East and the Central Plain.

Table 5.12 *Changing incidence of poverty in South East Asia, World Bank studies*[1]

	Date of survey[1]	Percentage of population below the poverty line			Percentage of population below the poverty line in 1985
		Total	Urban	Rural	
Indonesia	1976–79	47	28	51	39
Malaysia	1974–78	37	16	46	28
Philippines	1974–78	42	39	44	52
Thailand	1975–76	31	15	34	31

[1] There are serious problems of compatibility between these studies. For a detailed discussion see World Bank (1980) *Thailand: income growth and poverty alleviation*, Washington.

Source: 1976–9 World Bank (1985)

1985 World Bank (1988)

Table 5.13 *Changes in the incidence of poverty and the level of income inequality in Thailand, 1962–81*

Region and area	Poverty as % of population			%	Gini Coefficients[1]			
	1962/63	1968/69	1975/76	1981	1962/63	1968/69	1975/76	1981
North	65	36	35	23	0.359	0.370	0.422	0.456
Urban	56	19	31	23	0.460	0.440	0.453	0.462
Rural	66	37	36	23	0.308	0.345	0.368	0.422
North East	74	65	46	36	0.344	0.379	0.405	0.438
Urban	44	24	38	36	0.422	0.450	0.457	0.456
Rural	77	67	48	36	0.264	0.347	0.343	0.395
Central	40	16	16	16	0.391	0.401	0.399	0.430
Urban	40	14	20	24	0.384	0.399	0.425	0.455
Rural	40	16	15	14	0.375	0.392	0.376	0.418
South	44	38	33	21	0.402	0.401	0.449	0.456
Urban	35	24	29	18	0.360	0.450	0.465	0.443
Rural	46	40	35	22	0.370	0.325	0.402	0.426
Bangkok	28	11	12	4	n/a	0.412	0.398	0.405
Whole Kingdom	57	39	33	24	0.441	0.429	0.451	0.473
Urban	38	16	22	16	0.405	0.429	0.435	0.447
Rural	61	43	37	27	0.361	0.381	0.395	0.437

[1] The larger the Gini Coefficient, the more unequally income is distributed
Source: National Statistical Office Household Expenditure Surveys.

Table 5.14(a) *Thailand, Regional Per Capita Gross Product as a percentage of the National Per Capita GDP (at constant 1962 prices)*

	North East	North	South	Central[1] Plain	Central[2] Plain	Bangkok
1960	53.0	71.0	125.8	161.3	n/a	n/a
1961	53.4	70.0	119.2	165.5	n/a	n/a
1962	52.9	72.3	112.4	165.6	n/a	n/a
1963	53.1	71.2	110.5	167.8	n/a	n/a
1964	48.1	70.0	108.4	175.3	n/a	n/a
1965	50.0	69.3	108.4	175.3	n/a	n/a
1966	52.7	72.7	96.2	171.8	n/a	n/a
1967	46.9	71.0	97.1	181.0	n/a	n/a
1968	50.1	68.9	96.9	178.1	n/a	n/a
1969	48.5	68.9	98.2	179.0	n/a	n/a
1970[3]	45.0	67.4	96.2	183.0	n/a	n/a
1971	46.4	68.1	93.0	182.7	n/a	n/a
1972	48.3	67.8	92.4	181.5	n/a	n/a
1973	53.8	66.6	95.7	181.3	n/a	n/a
1974	45.0	63.0	100.7	184.1	147.7	274.1
1975	44.3	61.1	100.0	186.6	143.0	273.0
1976	44.0	61.7	100.3	191.0	148.6	268.3
1977	41.6	66.2	103.4	n/a	148.9	272.7
1978	40.4	65.8	104.9	n/a	148.5	273.2
1979	43.4	71.8	143.5	n/a	149.2	246.6

[1] Including Bangkok
[2] Excluding Bangkok
[3] In 1970 nine provinces were taken from the Central Plain and added to the North

Table 5.14 (b) *Thailand's Regional Per Capita Gross Product as a Percentage of National Per Capita GDP (at constant 1972 prices)*[1]

	North East	North	South	Central	Western	Eastern	Bangkok
1981	40.4	63.0	76.7	91.95	92.2	112.8	310.4
1982	41.9	64.1	77.8	92.2	105.2	120.6	293.7
1983	43.8	62.9	75.4	87.7	93.5	116.5	296.7
1984	43.2	65.4	74.1	93.8	96.8	116.2	290.5
1985	44.3	65.7	74.9	93.4	98.7	121.8	283.0
1986	39.2	63.3	74.7	90.0	97.7	123.9	290.5
1987	39.1	59.9	71.9	86.5	88.8	119.9	307.8

[1] These figures are not directly comparable to those contained in 5.14 (a) not only because of the different based year and the sub-division of the Central Region into Eastern, Western and Central but more significantly because of changes in the method of calculation (NESDB, 1989 *Gross Regional and Provincial Product 2524–2530*, Bangkok).

Source: National Economic and Social Development Board, Bangkok.

As has already been discussed (5.2), by far the greatest emphasis on rural development has been in Malaysia. Official statistics point to a sharp reduction in the incidence of poverty (table 5.15). However, as in Thailand this reduction may also have been accompanied by increased polarisation:

> in Malaysia economic development, considered praiseworthy in some regards, has not only maintained but probably even widened the income gaps within each major ethnic group in peninsular Malaysia, especially the Malays. (Sundaram, 1987: 114)

In an overview of agricultural development in Malaysia and Indonesia Gibbons *et al.* (1986: 206) concluded that not only did a high incidence of poverty persist but also that inequalities had increased:

> In absolute terms these inequalities are much greater in the Malaysian region where the spread of the new [agricultural] technology has been wider and of longer duration and where the agricultural growth has been greater. In other words development has been more unequal where agricultural growth has been greater.

Similarly, official data points to increasing regional income disparities (Malaysia, 1986, Fifth Plan: 170–3).

A major consequence of the concentration of development in the urban-industrial sector and the comparative neglect of rural development has been rapid and largely uncontrolled urban growth. As was discussed in chapter 1, South East Asia still has a low rate of urbanisation but an increasingly rapid rate of growth. Attempts to restrict and

Table 5.15 *Malaysia, changing incidence of poverty, 1976–1985*

	% of Households in poverty		Doctor per 10,000 population		Percentage of households with piped water		Percentage of households with electricity		Infant mortality rate per 1000 live births	
	1976	1984	1980	1985	1980	1985	1980	1985	1980	1985
Johor	29.0	12.2	2.1	2.7	48.8	69.0	40.3	62.9	20.0	17.3
Kedah	61.0	36.6	1.4	1.9	57.2	70.4	33.0	60.9	28.1	24.3
Kelantan	67.1	39.2	0.9	1.6	28.4	32.8	38.8	62.8	18.4	16.4
Melaka (Malacca)	32.4	15.8	2.5	3.3	76.5	84.4	54.2	70.9	18.7	20.3
Negri Sembilan	33.0	13.0	3.2	3.3	72.8	79.6	68.7	91.6	19.3	14.8
Pahang	38.9	15.7	1.9	2.2	58.8	72.8	34.0	49.5	15.9	13.7
Perak	43.0	20.3	1.9	2.7	68.2	82.4	25.3	58.0	19.3	21.3
Perlis	59.8	33.7	3.4	4.6	57.2	70.4	68.4	90.3	31.4	34.1
Pulau Pinang	32.4	13.4	2.2	3.7	87.0	91.0	70.8	81.0	23.5	23.2
Sabah	58.3	33.1	1.3	1.6	34.8	51.5	36.7	48.0	47.2	11.5
Sarawak	56.5	31.9	1.4	2.6	31.8	43.8	33.8	49.3	215.4	132.8
Selangor	22.9	8.6	2.2	2.9	80.8	88.1	44.8	81.0	27.5	27.0
Trengganu	60.3	28.9	1.2	1.7	46.4	59.1	49.8	70.6	14.7	12.2
Kuala Lumpur	9.0	4.9	10.2	11.4	80.0	88.1	71.4	95.7	20.7	12.6
Peninsular Malaysia	39.6	18.4	n/a	n/a	n/a	n/a	n/a	n/a	n/a	n/a
Malaysia	n/a	n/a	2.6	3.2	58.8	69.9	49.9	41.3	21.6	18.3

Source: *Fifth Malaysia Plan, 1986–1990*, 1986: 88 170–1, 510

control in-migration to urban areas have been limited and largely ineffective. Similarly, with the notable exception of Singapore, little progress has been made in planning for the growth of population in terms of housing, facilities or employment. The problems of congestion and growth that are not merely uncontrolled, but actually out of control, are seen at their most extreme in Jakarta. The most obvious consequences of this are the extensive squatter settlements, high level of un-and under-employment and dependence on often increasingly clandestine informal sector activities. Projections based on present trends suggest that between 1980 and 2000 the combined populations of Bangkok, Jakarta and Manila will almost double, reaching 9.5 million, 11.1 million and 12.0 million respectively.[42]

In line with the emphasis on attracting foreign investment and MNCs, all the South East Asian states have reorganised their capital cities to facilitate this. In many ways Singapore has acted as a model for these changes. The policies have these common characteristics: administrative reorganisation to enable the national government to assume direct control, in effect transforming the capital into a mini-state; major redevelopment and modernisation programmes; and suppression of the informal sector, particularly where it conflicts with modernisation (Drakakis-Smith and Rimmer, 1982). Overwhelmingly the redevelopment of the region's major urban centre has concentrated investment in the modern corporate sector – luxury hotels, apartments, offices, conference centres and mass transit systems – gauged to attract foreign investors rather than raise the living standards of the population.

Investment in low-cost housing, transport and facilities for the poor have been very much secondary considerations. The most spectacular developments have been in Singapore. However, while the population has almost certainly on balance gained from these developments the principal aims have been to establish a cheap, passive, easily controlled and stable workforce and a congenial environment for foreign investment and the operations of multi-national corporations.

All the countries in the region have established urban-development policies and planning structures. Frequently these have been part of broader national and regional programmes. In Indonesia, for example, the National Urban Development Strategy emphasises the importance of developing urban centres outside Java in order to promote a more spatially balanced pattern of economic development. However, for the capitalist countries of South East Asia:

> Any critical evaluation of policies adopted specifically for the purpose of influencing urbanisation is likely to reach the conclusion that these policies are relatively weak instruments compared with the basic and powerful effects of

certain underlying trends and of government policies adopted with other ends in view. (Jones, 1988: 145)

This statement could well have been made with respect to attempts to reduce the unevenness of development as a whole.

6

South East Asia in the late twentieth century: Problems and perspectives

6.1 Introduction

South East Asia's interaction with the wider regional and international economic structures has undergone frequent and profound change. Broadly these changes may be related to the successive emergence and dominance of mercantile, industrial and finance capital. South East Asia has successively developed as a supplier of high-value luxury produce such as spice; a producer of bulk primary produce, recipient of associated investment and as a market for mass-produced industrial goods; and as a major location for labour-intensive manufacturing operations and investment. Within the region these phases in the evolution of the world-economy have been reflected in the relations of production and the spatial pattern of economic activity. The progressive integration of South East Asia into the world-economy established the dominance of a small number of core areas and produced a pattern of increasingly uneven development at all levels.

The restructuring of the world-economy during the 1980s is beginning to open up South East Asia to a new cycle of capitalist penetration. Central to this process is the 'new orthodoxy' of development: the minimisation of barriers to foreign investment and MNC activity, the exploitation of 'comparative advantage'; and 'allocative efficiency'. These developments are in general contrary to the interests of the region's domestic capital. However, while in general the ability of the region's economies to resist the pressures for restructuring have been eroded, the process is operating extremely unevenly. This reflects the strength of domestic capital, the degree to which governments are unable to engage in economic restructuring without undermining their position, and the extent to which countries are considered to be economically and politically stable. Thus the renewed cycle of growth is far from evenly spread in the region, affecting principally Singapore and Thailand.

The rapidly changing world geo-political situation is likely to modify both the region's economic and strategic importance in the 1990s, and the distribution of activity within it. The prospects of China and Eastern Europe becoming more open to the activities of international capital and the associated lowering of international tension may well divert investment away from South East Asia and reduce the region's importance to, in particular, American military strategy. Within the region the now very real possibilities of Laos, Kampuchea, Vietnam and perhaps Burma re-engaging with international capital may well result in investment being diverted away from the presently pro-capitalist countries. This is most likely to affect those that are politically and economically unstable or where there is resistance to the activities of foreign capital (Chia Siow-Yue, 1986: 102).

It seems unlikely that restructuring, a new cycle of capitalist penetration or a changed world geo-political situation will radically alter the position that South East Asia has come to occupy in the world-economy. Current trends point to changes in degree rather than kind. The region may well become of lesser significance but it will remain locked into wider regional and international structures over which the countries, collectively or individually, have little control.

6.2 South East Asia and the Asian Regional Divisions of Labour

The capitalist states of South East Asia can be viewed at a number of levels. Firstly as part of the periphery of the world-economy with clear positions within both the old and new divisions of labour; secondly they are part of the Asian-Pacific region increasingly dominated by Japan; thirdly they are part of a wider East Asian economy centring on the four Asian NICs; and finally they can be seen as a regional economy pivoting on Singapore. It is on these latter two structures that this section focuses.

Singapore's emerging role as a regional and international business centre can only be understood in terms of the on-going process of global restructuring. Mirza (1986: 193–209) depicts Singapore as fulfilling a 'peripheral intermediate' role in the NIDL. That is, Singapore functions as a regional and global distribution centre for commodities and manufactured goods (an expansion of the entrepôt role), the East Asian 'inter-bank market' channelling excess funds between regions, and a centre for activities such as oil refining, producer services and manufacturing. Essentially Singapore's intermediate role in linking the resource-rich countries of South East Asia with the developed countries established as part of the British imperial structuring during the nineteenth century has been expanded into the financial and manufacturing spheres.

Within the South East Asian region, Singapore accounts for 8.5 per cent of intra-regional trade and 36 per cent of ASEAN trade with the rest of the world. Over 40 per cent of Singapore's trade is with other ASEAN countries (table 1.13). The main ASEAN exports to Singapore are oil and rubber and the main imports are chemicals, petroleum products, machinery and transport equipment. There is in this both a clear intermediate role and, increasingly, a regional division of labour.

MNCs establishing themselves in South East Asia have tended to establish their headquarters in Singapore. Thus much MNC investment in South East Asia is channelled through Singapore. In addition Singapore-based concerns and MNC branch plants are beginning to relocate labour-intensive production to other parts of the region where land and labour costs are lower while retaining the higher value-added process in Singapore.[1]

The developing links between Singapore and the adjacent Malaysian state of Johor illustrates the city state's evolving role in the region. Singapore-based firms (including MNC branch plants) operating labour-intensive processes are increasingly attracted to Johor where labour costs are approximately halved.[2] Similarly factory and office rents are much lower and power cheaper (although less reliable). At the end of 1986 there were 216 Singapore-owned branch plants in Johor mainly producing chemicals, furniture, electrical goods, textiles and rubber goods. These developments fit in with Lee Kuan Yew's vision of Singapore as the economic hub of South East Asia through which MNCs

> can serve not only their home market, but also the regional and world markets. Furthermore, they can tap the region's bigger and cheaper labour pool. Labour-intensive parts of a product can be done in neighbouring countries and exported to Singapore for the more capital- or skill-intensive operations. (speech given in Tokyo, cited Far Eastern Economic Review 18 August, 1989: 77)

The development of wafer-fabrication facilities in Singapore during 1988 further reflects the city-state's attempt to further South East Asian division of labour by making use of Thailand, Malaysia and the Philippines as cheap-labour locations for assembly work.

In this a clear exploitative intermediatory role can be seen; Singapore, having first partially lost its attraction as a low-cost labour centre and then completed the process as a matter of deliberate policy, is trying to insert itself between MNCs and cheaper labour locations.

While Singapore has a clear pivotal role within the South East Asia region, it has a broader function with respect to the Asian region. Henderson (1986; 1989) has illustrated this with respect to the electronics industry.

Following the establishment of semi-conductor assembly work in Hongkong by the US-based Fairchild Semi-conductor Corporation in 1972, the industry diffused into East and South East Asia laying the basis for an Asian Regional Division for Labour (ARDL). Hongkong and Singapore, as well as acting as regional headquarters for MNCs, also began to specialise in the assembly of small-batch, high-cost semi-conductors; the large-batch, lower-value assembly work moved initially to Malaysia, then Indonesia, Thailand and the Philippines (Siegal, 1980). These developments were accompanied by the establishment of testing facilities in Hongkong and Singapore for products assembled elsewhere. Since the early 1970s Motorola has used Hongkong as a testing centre for goods assembled in its Philippine, Malaysian and South Korean plants.

Essentially this regional division of labour has emerged in response to differential labour costs. Increases in wage levels have reduced the cost advantage of the Asian NICs for labour-intensive assembly work (table 6.1)

The intensification of these regional divisions of labour run counter to the stated aims of all the countries to move towards more sophisticated, capital-intensive production. In the electronics sector the low level of value added generated by semi-conductor assembly work has been a major incentive for countries attempting to move to more capital-intensive production. Value added is principally created in the design and fabrication stage not in assembly and testing.[3]

Malaysia is keen to move to greater sophistication in electronics principally because of fears of competition from Thailand. A number of new operations have been established in Thailand in response to the lower labour costs. Thus Malaysia is attempting to follow Singapore towards more capital-intensive 'higher-level' operations.

The more even spread of capital-intensive production rests in part on the flexibility of 'technical transfer'. Indeed a major element in the arguments of all the South East Asian government pronouncements in favour of foreign investment has been technology transfer. While transfer does take place all available evidence points to it being extremely limited in impact (Cunningham, 1983). As Sundaram (1987: 135) has noted:

Table 6.1 *Hourly compensations of production workers in the Asian electronics industry*

	1969[1] (US – 100)	1975[1] (US – 100)	1985[2] (US – 100)[3]	1985[2] Average hourly wage (US $)	Number of plants supplying 1985 data	Total respondents employment
Hong Kong	10	12	16	1.33	5	3749
Indonesia	n/a	5	4	0.35	1	1800
Korea	10	7	14	1.19	3	13073
Malaysia	n/a	9–10	10	0.84	6	11776
Philippines	n/a	6	8	0.63	8	11021
Singapore	9	12	19	1.58	6	4263
Taiwan	8	7	16	1.36	7	3196
Thailand	n/a	5	5	0.43	3	868

[1] Compensation in the international electronics industry
[2] Compensation in US-owned and locally-owned semiconductor assembly plants
[3] US standard of comparison given as $8.37

Source: Scott, 1987: 145

It is inconceivable that transnational corporations – reliant on technological superiority to ensure profitability – will voluntarily surrender technical edge, especially to potential competitors. This does not mean that no technical transfer can ever take place. Rather, such a transfer is planned to maximise profitability, not lose it. Hence, it would be naive to expect that such technology transfer can eventually develop an international competitive technology capacity.

In addition the practical training given to workers in assembly and testing work is highly specific and cannot be either translated into research and development or into other industries. The latter possibility is further limited by the generally low level of integration of the foreign-owned manufacturing plants with the rest of the economy.

The development of more capital-intensive production remains highly dependent on foreign investment, and thus on the confidence of investors and the absence of more attractive locations. Continued uncertainty over the political future of the Philippines is inhibiting foreign investment. Similarly in Malaysia foreign investment has declined, a reflection of investors' concern over the prospects of political and economic instability. This has been reinforced, as in Indonesia, by the continuation of politics which may restrict the activities of international capital. While Thailand has become a more attractive location for foreign investment, increasing attention is focusing on the possibilities of Laos, Kampuchea and Vietnam becoming open to international capital[4] (*Far Eastern Economic Review*, February 2, 1989: 12–13).

The Thai government, with its eyes set firmly on the goal of achieving NIC status during the 1990s, sees the former Indochinese states as playing a major role in this process.[5] This view is encapsulated in Prime Minister Chatichai Choonhavan's often quoted remark on the need to convert 'Indo-China from a battle-field to a market place'. The vaguely stated strategy envisages the Thai manufacturing sector utilising raw materials[6] from Laos and Vietnam and supplying them with manufactured goods. Thus a new mainland regional 'core' would result.

Given the deficiencies of Thai infra-structure and the poor linkages with the neighbouring states the prospect of Thailand becoming the 'Singapore of the mainland' remains open to serious question. While there is little doubt that Thailand would gain from the opening of Laos, Kampuchea and Vietnam to international capital it would also seem likely that, for example, Japanese and Taiwanese investment and products would enter directly through Vietnam rather than indirectly through Bangkok. However, a major influx of foreign capital into the former Indo-chinese states would radically alter the distribution of economic activity within the region as well as the South East Asian and Asian regional divisions of labour.

6.3 South East Asian states and the restructuring of the world-economy

In all the South East Asian economies the state has played a major role in the development of capitalist production. The relative weakness of the indigenous capitalist classes is reflected in the extent to which the military and the bureaucracy have come to exert decisive control over the state apparatuses. In general the policies of repression of labour organisations, subsidies on food and fuel, protectionism and investment in infra-structure and production have fostered the growth of the previously weakly developed indigenous capitalist class. In certain periods some of the states have been able to defend the interests of domestic capital against those of international capital. Complex inter-relationships have been built up between international and domestic capital, the bureaucracy and the military. This is perhaps most apparent in Indonesia and the Philippines. However, Robison's (1986: 374) description of the Indonesian situation could, with limited modification, be applied to much of the region:

> The state played its key role in the development of the capitalist class, not only providing the political conditions for capital accumulation, including the political repression of labour and the subsidisation of food and fuel prices, but actively investing in infrastructure and production. At another level it was active in resolving the conflicts internal to the class alliance, intervening decisively on behalf of domestic capital in the mid- and late 1970s. However, the very fragmentation of the capitalist class enabled the state to play a relatively autonomous role in its relations with capital. The relative weakness of indigenous elements within the capitalist class meant that no powerful bourgeois party emerged to challenge the formal hegemony over the state apparatus by military and, to a lesser extent, civilian politico-bureaucrats, whose political power grew out of the state apparatus itself. In this separation of political and economic power, the capitalist class, particularly international capital, financed the state and centres of politico-bureaucrat power which exercised hegemony over it, through formal revenues (oil taxes) and informal funding. The relationships between state, politico-bureaucrat and capital were further complicated by the bonds between individual politico-bureaucrats and capitalists, not to mention the fact that an increasing number of politico-bureaucrats were themselves also capitalists.

During the 1980s these complex interrelationships have been placed under increasing strain. Those that control the state apparatus have come to occupy a highly compromised position between the conflicting interests of domestic and international capital. Further compliance with the increasing demands of international capital threatens not only the political power base of the politico-bureaucratic group but also the stability of the state. Indeed there is an inherent contradiction between the policies pressed on the region's states by the IMF and the consequences of their implication for the interests of international capital.

These conflicts of interest are apparent in both the resistance by individual states to 'restructuring' and the 'softening' of the IMF demands. The latter has been most clearly apparent in the negotiations with the Philippines during 1988 and 1989.

Paradoxically those economies most attractive to international capital because of their relative political stability, resources, economic growth and market potential were often those which were best able to defend the interests of domestic capital. This was perhaps most clearly seen in Indonesia during the 1970s.

During the 1980s all the states have at various times and to different degrees compromised their domestic policy in favour of the interests of international capital. The situations that enabled countries to resist the pressure to abandon ISI in favour of EOI no longer prevail. In general, a combination of changed international circumstances and pressure from the international agencies has resulted in a sharp reduction in state-led development. The notable exceptions to this are Indonesia and Singapore.

State expenditure and budget deficits have been sharply curtailed, most spectacularly in Malaysia. There has been a general retreat from regional planning and controls over the economy. The 'buzz-words' of development are 'private sector', 'privatisation', 'allocative efficiency', 'free market' and 'comparative advantage'.

However, the rhetoric of development frequently remains separate from reality. Government and international agency statements cannot be taken at face value. The structures built up during the 1960s and 1970s have not and indeed cannot be removed at a stroke. Indeed as Robison *et al*. (1987, 11–12) have said, some states are 'caught in a bind' and, whatever the pressures, can realistically only engage in a very limited degree of restructuring. While it is tempting to compare the present partial adoption of the newest 'orthodoxy' to the transition from ISI to EOI discussed in 5.1 the situation in which this is taking place is a radically different one. For the South East Asian states the situation is much more precarious; their positions, internationally and internally, are weaker; the conflicts and contradictions of the development process are sharper; and the prospects of major political upheaval throughout the region greater perhaps than ever before.

6.4 Conclusion

Out of the process of global restructuring and the associated recurrent crisis of the 1970s and 1980s a new pattern of world production is beginning to emerge (Thrift, 1986a: 12). It may be still too early to infer the outcome of present trends. However, the world-economy is becoming more integrated at all scales. Most significant is the integration of international production and finance. As a result changes in the global pattern of capitalist production take place even more rapidly.

The key element in the emergent pattern is a new and more intense cycle of the internationalisation of capital. This process has operated extremely unevenly. The NIDL has only had a major impact on a small number of Third World countries and industries (Thrift, 1986a: 46–7). South East Asia is among the more significant to be so affected.

The new and intensified penetration of South East Asia was spearheaded by the influx of financial resources of which direct foreign investment is only the most obvious. The South East Asian states have been major recipients of the rapid expansion in foreign investment which has taken place since the early 1970s. Initially this was encouraged by their high levels of tariff protection. However, during the 1970s barriers to the operation of foreign investment began to emerge, most significantly in Indonesia and Malaysia. Restrictions on the free movement of foreign investment and MNC are now seen as barriers to the intensification of capitalist production in the region. Thus governments are under increasing pressure to dismantle the apparatus that was instrumental in the establishment of many of the region's manufacturing sectors.

Further development of South East Asia's position in the NIDL depends, of course, on the continuation of the trends giving rise to this new global pattern. As Jenkins (1984) has suggested the NIDL as we currently recognise it may prove to be a very temporary phenomenon. Moves towards automation of labour-intensive processes, reinforced by increased political uncertainty in South East Asia and increased protection of developed world markets, may well result in a movement of MNC activity and foreign investment back to the USA and Western Europe. The major American-based electronics MNCs, such as Motorola and Fairchild, are beginning to repatriate assembly and testing work back to the USA (Henderson, 1986: 104).[7] Indeed a study by OECD (1988) suggests that this may become a major trend as the NICs as a whole lose their cost advantages over the developed economies.

The growing tendency for Japanese manufacturing investment to flow to developed rather than less-developed locations is principally a reflection of growing protectionism. With the prospects of the world-economy moving towards 'highly managed regional trading blocs' (GATT, 1988), Hongkong, Singapore, South Korea and Taiwan are beginning to follow the Japanese lead. It is likely that these flows will increase with, for example, the prospect of increased EC protection after 1992 (Wilson, 1988).

If the currently rapid opening of China to international capital continues it is likely that much activity will be diverted away from South East Asia. The increasing likelihood of opportunities opening up in the USSR and Eastern Europe can be viewed in a similar way.

This concluding discussion highlights the need for an international perspective. The degree of integration and speed of change exhibited by the world-economy make it increasingly difficult to produce a meaningful analysis of individual countries or regions such as South East Asia in isolation. Rather they must be studied as interacting parts of the world-economy.

Notes

1 Contemporary South East Asia in the world-economy

1 For a discussion of the name 'South East Asia' see Emmerson (1984).
2 See Fisher (1964: 3–10) for a discussion of these terms.
3 The emergence of these new structures is discussed in chapter 6. For alternative ways of placing the South East Asian states in broader regionalisations of the Asian Pacific Region see Forbes *et al.* (1985) and Berger and Hsiao (1988).
4 In addition to tin, South East Asia produces a wide range of minerals including coal, copper, chromium, gemstones, lead, silver, rare-earth metals, tungsten, titanium and zinc. Few of these are produced in sufficient quantity to be of major significance to the world-economy. However, Burma produces 2 per cent of world tungsten; Malaysia 11.6 per cent of rare-earth metals and 7 per cent of titanium; and Vietnam 2 per cent of anthracite.
5 The hydro-carbon potential of South East Asia is estimated at 180,000 million barrels of oil and over 3,420,000 million cubic metres of recoverable natural gas. These represent 60 per cent of the reserves of the Asia-Pacific region and 9 per cent of world reserves (Shoebridge, 1988: 66–7). Production is dominated by Indonesia (table 1.4b) which exports nearly 70 per cent of production and is now the world's largest exporter of liquid natural gas. In Thailand and Vietnam the reserves suggest that there is considerable potential for increased self-sufficiency, while the Philippines are likely to remain heavily dependent on imports.
6 For a discussion of the Indonesian and Malaysian views see Tjeng (1985) and Jeshurun (1985).
7 These developments were effectively brought to an end by the reaction of ASEAN to the Vietnamese 'invasion' of Kampuchea.
8 For a discussion of the possibilities of a Vietnamese invasion of Thailand and related threats to South East Asian security see Pike (1985: 186–8).
9 For example the USSR is depicted as having the ability to control the South China Sea and the Straits of Malacca, thus severing Japan's oil supply routes.

10 Cho (1990) suggests 'that the close-knit relationships among the various Royal Houses and the top echelons of the Armed Forces, the strong sense of loyalty towards the Sultanates and the symbolic ranks which Sultans traditionally have as Commanders-in-Chief in the Armed Forces may provide some of the reasons for the lack of direct involvement in government'.
11 These may well be serious underestimates.
12 See section 5.2, particularly footnotes 26 and 27, for a discussion of difficulties relating to Indonesian foreign investment.
13 For a discussion of the decline of EC trade with South East Asia see Taylor and Ayre (1986).
14 In 1986 24 per cent of Laotian exports were recorded as 'manufactured goods', but this comprised almost entirely hydro-electric power.
15 In mainland South East Asia illegal trade flourishes. Consumer goods from Thailand are smuggled into Burma, Laos and Kampuchea. In return tobacco, silk, precious stones, cattle and teak enter Thailand. This is in addition to, though sometimes connected with, the drug trade.
16 Estimated by Kraus and Lütkenhorst (1986: 36).
17 Coefficients of Variability for the ASEAN per capita GDP rose from 71.1 per cent in 1967, to 108.5 per cent in 1977 and to 116.2 per cent in 1987.
18 The contribution of agriculture to GDP is generally underestimated because:
   production is under-recorded
   the subsistence element is under-valued
   the movement of the sectoral and international terms of trade against agriculture.
19 During the 1970s South East Asia had a rate of population growth that was remarkably high by Third World standards (see Dixon, 1990b for a discussion).
20 See Griffith-Jones (1984) for a discussion of 'debt-led growth' in the Asian economies.

2 Pre-Colonial South East Asia

1 Roberts (1985: 310) notes that:

> The stereotype of the 'unchanging East' made its appearance in the eighteenth century. In a culture [that of England] more and more aware of its own capacity for change; and change for the better, it led easily to the dismissal of Asiatic civilizations as 'stagnant' ... by consequence of bad government, the natural depravity of their inhabitants or many other causes.

As illustration of this he cites Leroy-Beaulieu (1891: 165) who said that the world

> is composed of four different parts, in terms of civilisation. That of western civilisation – our own part. A second part is inhabited by people of a different civilisation but organised in coherent and stable societies and destined by their history and present character to govern themselves – the Chinese and Japanese people, for example. In the third part live people advanced enough in some respects, but ones

which have either stagnated or have not been able to constitute themselves as unified, peaceful, progressive nations, following a regular development ... India ... before the British conquest, Java and the Indochinese peninsula present particularly this third type. Finally, a great part of the world is inhabited by barbarian tribes or savages, some given over to wars without end and to brutal customs, and others knowing so little of the arts and being so little accustomed to work and invention that they do not know how to exploit their land and its natural riches. They live in little groups, impoverished and scattered, in enormous territories which could nourish vast numbers of people with ease ... This state of the world implies for the civilised people a right of intervention ... in the affairs of the last two categories.

2 There is still a considerable amount of widely available literature that throws doubt on the ability of South East Asia to have produced any sustainable level of 'higher civilization':

> In the writer's opinion this decline of Indianized civilization in South East Asia would seem to provide substantial confirmation of Gourou's [expounded in *The Tropical World*, 1953] thesis that the highest forms of civilization are not indigenous to the tropics and can be maintained there only as a result of vigour imported from outside. Thus ... the great flowering of civilization in the humid tropical fringe of South and South-east Asia during the millenium which preceded the impact of the West may not unreasonably be viewed as the outcome of such a tide which, after sweeping over the Indian subcontinent from the north-west, ultimately petered out in ripples of ever diminishing intensity across the peninsula and archipelagos beyond.' (Fisher, 1964: 94)

3 It has perhaps the greatest structural complexity of any similar sized region (Fryer, 1979: 36).

4 As with much pre-colonial South East Asia the evidence relating to Srivijaya is extremely limited. The location, extent and dating of the state remain areas of considerable debate. For a recent review see Andaya and Andaya. (1982: 17–36).

5 With the exception of the northern part of the region, the mean annual sea-level temperatures are remarkably close to 27°C in Singapore and up to 11°C in, for example, Mandalay.

For a detailed description of the climate of South East Asia see Fisher (1964: 21–42).

6 South East Asia may well have been the region of origin for *oryza* (Vavilov, 1926). More recent work (Solheim, 1972; Gorman, 1977; Higham, 1984) suggests that wet rice was cultivated in mainland South East Asia as early as 5,000–7,000 BC.

7 For a discussion of the relative productivity of wet rice and dry crops see Bray (1986: 15) and Farris (1985: 17).

8 The ability to remove water from the fields can be as important as irrigation. Indeed in the lower valleys of the Red River, Mekong, Irrawaddy and Chao Phraya flood control is vital for reliable rice cultivation.

9 This does not imply in any way acceptance of the Wittfogel (1957) thesis of 'hydraulic society'.

10 While the core of the Khmer Empire comprised extensive irrigated wet-rice

cultivation, a considerable proportion of production almost certainly came from unintensive cultivation and indeed from shifting cultivation (Fisher, 1964: 115, citing Pendleton, 1962).

11 Fisher (1971: 87) estimated that the Western two-thirds of South East Asia was directly influenced. However, more recent studies (for example, Hall, 1982) suggest that the impact and extent of Indian influence in South East Asia has been exaggerated.

12 The Thai and Khmer scripts, for example, are based on that of Sanskrit.

13 For a fuller summary see Krader (1981).

14 For a brief resume of Onghokham and other work questioning the validity of the view that pre-colonial Java was dominated by communal village-based production see Warren (1985: 130–1 and footnotes 9 and 10: 141). It should be noted that Geertz relied heavily on nineteenth-century Dutch accounts and on the earlier work of Boeke (1947; 1953).

15 A reference to Tawney's [1966: 77] oft-quoted description of rural China in 1931:

> There are districts in which the position of the rural population is that of a man standing permanently up to his neck in water, so that even a ripple is sufficient to drown him.

16 In the pre-colonial period a wide range of new subsistence crops – sweet potatoes, cassava and maize – introduced by the Portuguese from South America spread through the region. Similarly new varieties of rice, methods of cultivation and water control were adopted, frequently imported from China and India.

17 See Ingram (1971: 29–33) for a summary of estimates of Thai state revenues.

18 Some estimates are as high as 350,000 (Sternstein, 1984).

19 This account is drawn from Oki (1984: 270).

20 This discussion of land-holding and labour obligations is based on Wales (1934), Grahame (1924) and Ingram (1971: 12–13); Hong Lysa (1984).

21 These were not simply a hereditary nobility; there was considerable social mobility within the ruling class. Royal titles were lost at the fifth generation and royal grants of land ennobled functionaries (Elliott, 1978: 45).

22 See Murray (1980: 49–52) for an account of the development of cash rents in Vietnam.

3 Western penetration: from trade to colonial annexation

1 Braudel (1979, volume 3: 484) refers to 'the Far East as the greatest of all the world-economies' which:

> taken as a whole, consisted of three gigantic world-economies: Islam, overlooking the Indian Ocean from the Red Sea and the Persian Gulf, and controlling the endless chain of deserts stretching across Asia from Arabia to China; India, whose influence extended throughout the Indian Ocean, both east and west of Cape Comorin; and China, at once a great territorial power – striking deep into the heart of Asia – and a maritime force, controlling the seas and countries bordering the Pacific. And so it had been for many hundreds of years.

But between the fifteenth and eighteenth centuries, it is perhaps permissible to talk of a *single* world-economy broadly embracing all three.

2 'Spice' was a very broad term which included drugs, resins and glues as well as condiments.
3 This was related to increased buying power and increased consumption of preserved meat.
4 The only other viable route was through the Sunda Straits, a much more hazardous route in navigational terms. However, until *c.*1400 the Gujerati spice traders used the Sunda Straits in a route that ran along the west coast of Sumatra, through the Sunda Straits, along the north coast of Java. This route probably resulted from their original concentration on trade with the ports of north-eastern Sumatra – Aceh, Pedir and Pasai, and a desire to avoid piracy in the Straits of Singapore and the Riau Archipelago. During the fifteenth century Gujerati and Bengali trade passed increasingly through the Straits of Malacca (Harrison, 1963: 52).
5 The town was founded in 1403 and by 1409 was playing a major role in trade (Braudel, 1979: 524). For a concise account of the locational advantages of Malacca see Andaya and Andaya (1982: 39–44). See Subrahmanyam (1988) for summaries of several contemporary descriptions of Malacca during the early 1500s.
6 Tomé Pires' *Suma Oriental* is probably the most important and complete account of South East Asia in the early years of the sixteenth century. Pires arrived in India in 1511. Between 1513 and 1517 he was based in Malacca where the work was completed. In 1517 he became the first Portuguese ambassador to China. He appears to have visited many parts of island South East Asia and possibly Cochinchina, Siam and Cambodia. Most of the secondary references used in the present chapter appear to have drawn heavily on this work.
7 For a discussion of the evidence see Reid (1980). During the period few European cities except Paris and Naples exceeded 40,000.
8 See, for example, Brenner (1977), Wallerstein (1974) and Braudel (1979).
9 In terms of land-based military power the Portuguese were in no position to confront the Asian states and indeed were generally obliged to deal with them as they would other European powers. However, at sea the situation was very different. The navigational conditions of the Atlantic had given rise to hull and sail designs that were capable of much greater manoeuvrability than shipping that had evolved to cope with the monsoonal system. In addition European ships had begun to mount artillery. A series of decisive naval engagements, most notably the battle of Diú in 1509, reflects the advantages that the Portuguese had at sea. For a detailed account see Cipolla (1965).
10 The Portuguese were never able to suppress the overland trade in spice, despite their presence at Ormutz. Indeed the trade continued to flourish for a hundred years after the opening of the sea route. In part the quality of spice may have been higher – the storage conditions on the voyage around the

Cape of Good Hope must have been far from ideal. Wallerstein (1974: 340) cites material that suggests that the Portuguese also purchased inferior spice.

11 For a detailed discussion of the returns on the spice trade see Godinho (1969: 683–709), who cites examples of spice cargoes selling for eight times their purchase price. However, the costs of the round trip, and shipping losses of almost 11 per cent on the outward voyage and almost 15 per cent on the return have to be considered. (cited Wallerstein, 1974: 340).

12 The most comprehensive and probably most reliable source is Tomé Pires (see footnote 6).

13 Mexican silver dollars went in such large quantities to Canton, Amoy that they became the medium of exchange for Far Eastern international trade (Hall, 1981: 274).

14 The Philippines were on the fringe of the main cultural movements of Asia. Even the largest islands comprised a mosaic of ethnic and linguistic groups.

15 Augustinians, Franciscans, Dominicans, Benedictines, Recollects and Jesuits. The islands were divided into areas of activity for each of these orders (see Tate, 1971: 343, figure 46).

16 The active spread of Islam in island South East Asia may well have been accentuated by the Portuguese activities. Certainly Islam came to be a focal point for anti-Portuguese interests throughout the islands (Tate, 1971: 45).

17 For a discussion see Wallerstein (1974) and Brenner (1977).

18 'Dutch expansion to the East formed a major item in their eighty years' struggle for independence and was undertaken as much for political and strategic as for economic reasons. Their East India Company conducted a concentrated national offensive against Portugal and Spain' (Hall, 1981: 335).

19 Initially the route enabled traders to avoid the Portuguese-dominated Straits of Malacca. The use of the more direct route from the Cape of Good Hope (map 3.3) reflected the advances made in navigation and shipbuilding.

20 As with the Portuguese the spread of Islam acted as a focus of opposition to the Dutch. See Schrieké (1929) for a discussion of the role of Islam in providing an ideological basis for opposition to the Dutch.

21 This was part of the wider conflict around the American War of Independence, 1776–84. This involved France from 1778, Spain from 1779 and Holland from 1780.

22 A number of short-lived settlements were established, for example on the islands of Balambangan and Labuan between 1771 and 1775. It was also considered by the EIC and the Admiralty that a base to the east of the Bay of Bengal was of strategic importance during the period of the north-east monsoon.

23 British forces occupied Malacca in 1795, Ternate in 1801 and Java in 1811.

24 In the 1590s a group of Portuguese adventurers attempted to take over Cambodia (see Tate, 1971: 473 and 478 for a concise account).

25 As Tate (1971: 503) notes, the attempt was premature. Neither France nor the other European powers had the capacity to annex the mainland states in this period. For an account of this extraordinary affair see Hall (1981: 279–83).

26 It was not merely the weapons that were imported but also the techniques of their manufacture.
27 Hybrid junk-brig designs that rivalled the largest shipping used by the EIC were emerging from the Thai yards (Sarasin Viraphol, 1977).
28 There was some very direct involvement in internal affairs. For example the presence of French 'volunteers' in Cochin China in the 1780s and 1790s played some part in the emergence of the Vietnamese state (Hall, 1981: 453).
29 This increased from £1.3 million in 1792 to over £10.5 million in 1807 (Sar Desai, 1977: 25).
30 Penang lacked the 'nodality' of Singapore for the regional trade. In addition shipping was vulnerable to pirate attack during its long haul against contrary winds up the Straits of Malacca (Tate, 1979: 163).
31 By 1824 the trade was valued at over $11.4 million; 26 per cent of this was with the islands and 13 per cent with the Malay Peninsula (Sar Desai, 1977: 43).
32 The further opening of the China trade after 1858 was matched by increased 'direct shipment' which bypassed Singapore. In addition the French and Dutch territorial expansion was increasingly effective in excluding British traders. For a summary account of these changes see Sar Desai (1977: 140–8).
33 The pressure mounted by the business community was reflected in the *Penang Gazette*:

> the relations of England to the native states of Malayan Peninsula require revision ... The riches of the peninsula must not and cannot remain longer undeveloped, but must be made to contribute to the wealth of the world ... In plain words these states must be brought under European control. (Penang Gazette, 12 July 1871, cited Musimgrafik, 1979: 35)

34 The strategic importance of the Arakan coast for the defence of India had become apparent during the wars of 1778–84, 1793–1803 and 1803–15.
35 The expansion and unification of Burma during the late eighteenth and early ninteenth centuries, however, drained the kingdom of resources.
36 This section relies heavily on Hall (1981).
37 Spain participated ostensibly because of the persecution of Jesuit missionaries.
38 This area was usually referred to as Siem Reap.
39 The degree of official French involvement in the affair is far from clear. Francis Garnier had been sent to Hanoi with a small force (*c.*121 troops) to arbitrate in a dispute between a French trader and Vietnamese officials. Garnier apparently exceeded his instructions, taking advantage of the disrupted conditions in Tonkin to seize the citadel in Hanoi and the fortifications of Tourane. While Paris repudiated this action it led directly to the 1874 treaty, the failure to honour which legitimised subsequent direct intervention.
40 The difficulty of obtaining profitable returns for our imports has always acted injuriously on our trade with Siam; but now that the staple export of the country is monopolized contrary to Treaty [Burney treaty of 1826] the difficulty is very greatly

> increased and this coupled with the exorbitant dues levied on British shipping has reduced our trade to one-tenth its natural dimensions.
>
> Merchant's memorial 1849 cited by Frankfurter (1904: 196).

41 Mongkut had spent 20 years in the monkhood and had engaged in extensive studies of western history and literature. He appears to have been very aware of the pressures on Thailand to open its doors to the West, the dangers of resistance and the possible benefits that might result.

42 As Cady (1958: 143 and 148) has pointed out, the signing of the treaty took place against the background of the British annexation of Lower Burma and the presence of a British naval force.

43 By the 1930s 70 per cent of Thai trade was with Britain and 95 per cent of the 'modern economic' sector was foreign owned, mainly by Britain.

44 Quoted by Steinberg (1987: 159).

45 This has been described as 'three hundred years in a Jesuit seminary and fifty years in Hollywood'.

4 Uneven development: the establishment of capitalist production

1 The Dutch term *cultuurstelsel* is more accurately translated as 'cultivation system'.

2 Chin Yong Fong (1977: 124) suggests that at its peak 50 per cent of the Javanese population was affected. While Geertz (1963: 53) considered that the 'culture system' was a decisive factor in the development of the present pattern in Indonesia he also stressed that it:

> was only part of a much larger complex of politico-economic policies and institutions. Alongside the mammoth state plantation which this system tended to make of Java, there was a whole series of adjuncts, related systems, and independent growths, so that the picture of the island from 1830 to 1870 is a much more differentiated and much less static one that than which has so often been drawn for us.

A number of studies have confirmed the existence of divergent trends of development – for example, Fernando (1988) who, for part of West Java, traced a more 'conventional' pattern of commercialisation with marked differentiation of the community into large landowners and landless labourers.

3 Ingram (1971: 196) described the adviser as 'one of the most powerful figures in the government until about 1930. Without his approval a foreign loan could probably not have been marketed successfully, and the government could scarcely have carried out any financial measures to which he had strong objections.' Brown (Brown, 1978: 210) has seriously questioned this view, concluding 'that the British Financial Advisers had relatively little effective influence in Siam in the late nineteenth and early twentieth centuries'.

4 From a memorandum of 1908 to the Minister of Finance from the Financial Adviser, W.J.E. Williams, headed 'Further proposal schemes of irrigation development' (cited by Ingram, 1971: 198).

5 Chinese hydraulic mining and smelting techniques were superior to those of the Malays.
6 Straits Settlement tin was of a finer quality than that of Cornwall.
7 These are Parish's (1914: vi) figures; for a discussion of their derivation and accuracy, see Platt (1986: 87–110).
8 This is the total for the period 1880–1918.
9 See, however, Davenport-Hines and Jones (1989).
10 In 1896 when he became the Resident-General of the Federated Malay States he reiterated this view, stressing that it was the duty of the Colonial Administration 'to open up the country by great works: roads, railways, telegraphs, wharves' (cited Chai Hon Chan, 1967: 66).
11 For example, the 1885 legislation in the NEI which was aimed at the establishment of individual ownership of land and the breakdown of communal production (Furnivall, 1944: 319).
12 These were referred to as the 'three beasts of burden'.
13 Dunn (1928: 20) cited by Adas (1974: 144).
14 The export price of rice at Bangkok, for example, fell from 117.4 baht/ton in 1928–9 to 48.3 in 1934–5 (Feeney, 1982: 128–9).
15 Furnivall (1938: 74) refers to this as 'industrial agriculture' because of the division of labour between the gangs.
16 There had been increasing pressure by Dutch capitalists for an end to the 'Culture System' (Tate, 1979: 54–5).
17 By 1938 there were 2,400 estates, the majority controlled by a few large interlocking companies (Geertz, 1963: 86 and Wertheim, 1959: 99).
18 Some planters and capital moved from Ceylon where fungus infection had destroyed the industry after 1869.
19 European estates were normally between 400 and 800 hectares. During the 1930s they tended to concentrate into units of over 2,000 hectares (Allen and Donithorne, 1957: 129). In contrast, Chinese production was concentrated in estates of 20–200 hectares (Fisher, 1964: 611).
20 The extent of this sector is difficult to gauge because official statistics classify 'small-holders' as less than 40 hectares, a size that included many specialised Chinese estates.
21 See Murray (1980: 241–6) for an account of labour recruitment.
22 For a detailed account of the development of craft industries see Gourou (1936: 460–505).
23 In 1902 the US Congress extended the Chinese exclusion laws to the Philippines.
24 By the 1930s the most diversive mineral sector was that of Burma:

	1939
Burma proper	
Tin concentrates, tons	5,441
Tungsten concentrates, tons	4,342
Antimony, tons	345
Rubies and sapphires, carats	222,102

Jadeite, cwt	767
Gold, oz	1,206
Shan State	
Silver, 000 oz	6,175
Bawdwin	
Lead and concentrates, tons	77,180
Zinc, tons	59,347
Copper matte, tons	7,935
Nickel spiers, tons	2,896
Iron ore, tons	26,259
Karenni	
Tin and tungsten concentrates, tons	5,593
Burma	
Petroleum, tons	1,064,376

Source: Stamp 1956: 398

Small and less diverse mineral sectors had developed elsewhere in the region notably in French Indo-China:

Mineral	Production in thousands of piastres (1937)	% of total
Coal	12,105	63
Tin and tungsten	5,689	29
Gold	512	3
Zinc	384	2
Others	580	3
	19,270	100

Source: Guillaumat, 1938: 1250

and the NEI:

	1937	
Coal	1,754,000	(tons)
Bauxite	273,877	(tons)
Gold and silver	15,555	(kg)
Petroleum	59,109,000	(barrels)
Tin	15,555	(tons)

Source: Naval Intelligence Division, 1944: 247–70

In the Philippines minerals were of little importance with only limited development of lignite, copper, gold and iron production during the 1930s.

25 The enthusiasm for the Chinese exhibited by Raffles and many subsequent British administrators was shared, though generally to a lesser extent, by the Dutch and French representatives.

26 Restrictions on immigration were imposed by the American authorities after 1902 (Purcell, 1965: 620).

27 During the early nineteenth century these fields were producing one-seventh of world supply (Crawfurd, 1828, vol. 3: 486).

28 The most graphic description of this traffic is contained in Joseph Conrad's novel *Typhoon*.
29 The communities were by no means homogenous, being drawn from a wide area of southern China, in particular the province of Kwangsi, Kwangtung, Kiangsi, Fukien, Foochow and Hainan. Migrants spoke a variety of languages, notably Hokkien, Teachiu, Cantonese and Kakka.
30 Similarly, by the late 1930s by far the largest groups of inhabitants in Rangoon were Indian.
31 The 1891 census of the Straits Settlements and the Federated States reported 12.7 per cent of the population as Indian, 43 per cent as Chinese and 44.3 per cent as Malay (cited Ooi Jin Bee, 1976: 113). There was little change in these proportions during the remainder of the colonial period. In 1941 14 per cent of the population were Indian, 44 per cent Chinese and 42 per cent Malay (Ooi Jin Bee, 1976: 122).
32 By 1936 the Indian share of total investment in Burma was 62.1 per cent.
33 If the Netherland East Indies are excluded, Chinese and Indian investment represented 53.8 per cent of the regional total by the late 1930s.
34 In both Java and Indo-China 'craft' products were developed into a thriving export trade.
35 In Indo-China the recruitment of labour for 'war work' in France made a lasting impression on the organisation and political awareness of the working class.
36 In 1937 it has been estimated (Gould, 1950) that 110,000 workers were employed in the manufacturing sector, representing 1.6 per cent of the working population and contributing 9.9 per cent of national income. However, Ingram (1971: 144) considers that these figures include a considerable amount of primary processing. Similarly in Burma, while the 1931 census records 560,000 people, 10.7 per cent of the working population, engaged in manufacturing, Hill (1984) argues that only 90,000, 1.5 per cent of the working population, were employed in factories. For the other countries figures cited included:

   NEI *c.* 150,000 (*Naval Intelligence Division*, 1944: 273).
   French Indo-China *c.* 140,000 (*Naval Intelligence Division*, 1943: 321).

37 During the Japanese occupation local capital moved into a variety of new areas in order to meet demands no longer met by imports and to serve the Japanese war effort.
38 Within the region official contact between the territories of rival colonial powers was minimal. These divisions were strengthened by the imposition of Dutch, French, English or Spanish as the 'official' language.
39 While the trade of the various South East Asian territories became firmly locked into increasingly rigid imperial trading structures a degree of intra-regional trade continued, largely in the hands of the Chinese (Fisher, 1964: 186; Wong Lin Ken, 1978: 50). Regional and extra-regional exchange were linked through the largely European-controlled agency houses. These were increasingly based at Singapore which came to provide the vital 'interface'

between regional trade and production, and international trade and finance (Khoo Kay Kim, 1972: 95–100). The port-city began to acquire an intermediary role in the Western trading structures as well as becoming a port of call on the major sea routes from the Far East and Australia to India and Europe.

40 For an excellent review see Owen (1987a).

41 For a series of studies which indicate the limited impact of health measures in the region see Owen (1987b).

42 Bangkok was founded in 1782.

43 Thailand was by far the most extreme case. By 1930 Bangkok was fifteen times the size of Chiangmai (Sternstein, 1984: 67).

44 The ratio of Saigon-Cholon's population to that of Hanoi's fell from 3.3 per cent in 1921 to 1.8 per cent in 1939 (McGee, 1967: 54).

45 Though by far the most organised.

46 In 1930 the number of plantation workers in Cochinchina was officially recorded as 49,000. However, Bunout (1936: 112) considered the true figure to be nearer 100,000 (for a discussion see Murray, 1980: 222).

47 See Brown (1986).

48 It was frequently asserted that under the Culture System the Dutch grew in wealth and the Javanese in numbers (cited Geertz, 1963: 70).

49 The state-promoted credit schemes did not end rural indebtedness and, in effect, only supplemented the established moneylenders (Tate, 1979: 101).

50 In 1941 41,000 migrants left Java, but this only represented 7 per cent of that year's population increase (Tempelman, 1977: 16).

51 For a review see Brown (1986; 1988).

52 Cited Khoo Kay Kim, 1977: 86.

53 Cited Khoo Kay Kim, 1977: 86–7.

54 In the absence of significant state welfare programmes a wide range of volunteer organisations were established, for example the Destitute Chinese Emergency Fund and the Selangor Indian Association. Some of these developed into important political organisations (see Khoo Kim Kay, 1977 and Cheah Boon Kheng, 1977).

55 Brown (1986: 1010) questions Scott's conclusions, suggesting that they should be 'treated with some caution'.

56 In the Philippines opposition to Spanish rule gave rise to the Propaganda Movement of the 1870s, the Liga Filipina organisation in 1892 and in 1899 the declaration of the First Philippine Republic. Following the defeat of the Republic and the imposition of American rule the independence movements continued to develop. Their main period of expansion was the 1920s and 1930s (Ongkili, 1977: 232).

57 The 1906 Village Act was intended to establish some limited form of local responsibility. In 1916 a People's Council was established. Until 1928, this essentially advisory body comprised mainly Dutch and nominated members. The 1925 constitution led, in 1929, to the twenty-two regencies being combined into three provinces. In these the governor was assisted by a partly

elected council with a non-European majority (see Hall, 1981: 790 for a further discussion). None of these developments gave any real power to non-Europeans.

58 Hall (1981: 800) refers to the 'Consultative Native Assemblies' as 'an excellent example of the camouflage system used by the French'.

59 From 1867 until 1937 British territory in Burma was administered as an Indian province. Indeed it appeared on maps of the period as 'Further India'.

60 For a detailed discussion of this in the context of wartime planning for the reoccupation and development of the Burmese economy see Tarling 1978(a) and 1978(b).

61 The only real involvement was of Indians in low-level administrative positions (Golay *et al.*, 1969: 348).

62 During the Japanese occupation the position of the traditional rulers was weakened and the nationalism of the Malay elite that had emerged during the 1920s and 1930s was encouraged. The Indian and Chinese populations were discriminated against, the latter often in a most brutal manner.

The focus of opposition to the Japanese was the MPAJA (Malayan People's Anti-Japanese Army) which was predominantly Chinese and largely under the control of the MCP (Malayan Communist Party). Experience gained in the resistance to the Japanese was carried over into MCP activity during the emergency.

For an account of the development of political groupings and the growth of nationalism during the occupation see Andaya and Andaya, 1982: 247–54.

63 Throughout Asia the Japanese encouraged the independence movements and engaged in widespread anti-Western propaganda, centering on the slogan 'Asia for the Asians'. Cynical observers saw this as meaning 'Asia for the Japanese'.

64 Fisher (1964: 184) gives the following figures for Japanese:

Burma	465	(1931)
Thailand	522	(1937)
Indo-China	241	(1936)
Malaya	8,000	(1931)
Indonesia	6,500	(1930)
Philippines	29,000	(1938)

For a review see 'Japan and the Western Powers in Southeast Asia', special issue of *Journal of Southeast Asian Studies*, 9, 2, 1978.

65 The USA signed a treaty in 1920 which recognised Thai fiscal autonomy. In practice this had little immediate impact because of the dominant position of the UK. However, the USA was vocal in its support of Thai appeals for treaty revision. Britain strongly resisted these demands until 1926. The 1926 agreement, while giving Thailand fiscal autonomy, retained preferential duties on cotton goods, iron and steel goods and manufactured items – Britain's principal exports to the Kingdom – until 1936.

66 The expansion of tin and rubber exports to the USA and Japan were assisted by the generous quotes that the Kingdom had extracted as a price for joining the Anglo-Dutch inspired tin and rubber restriction schemes (Caldwell, 1978: 10).

67 The 1926 treaty revision allowed Thailand to raise import tariffs. Duty on imported cigarettes was set at 25 per cent in 1926, increasing to 50 per cent in 1931 (Ingram, 1971: 136).

68 The impact of the occupation was in these respects extremely uneven (see McCoy, 1980).

69 In many instances nationalist groups that had welcomed the Japanese as 'liberators' changed their opinion as a result of the occupation. However, their anti-Japanese stance did not mean that they were in favour of a reimposition of Western colonial rule.

70 In 1948 the government controlled little of the country beyond the immediate Rangoon area.

71 While the Dutch were able to hold the urban areas, they were unable to effectively control the countryside.

72 See Palmier (1962) for a detailed account of the international pressure and the resistance of the nationalists to Dutch ambitions. As Mohammad Natsir, the Minister of Information stated: 'the big cities can be occupied by Dutch forces, but the countryside cannot be captured. The Dutch lack the power to dominate the whole of Republican territory, and they lack the money to maintain thousands of soldiers . . .' (cited Kahin, 1952: 394).

73 Shortly after independence the city of Bandung was seized by an irregular force of Indonesians and Europeans commanded by Captain Raymond Westerling, a Netherlands citizen. This force made an unsuccessful attempt to gain control of West Java, some units penetrating Jakarta. Following defeat by the Federal forces Westerling escaped to Singapore in a Dutch aircraft. While there is reason to doubt that there was official Dutch involvement in this event, anti-Dutch sentiments were inflamed. For an account of this and other events in the deterioration of the Dutch-Indonesian relations see Palmier, 1962: 75–137.

74 For a summary of these see Smith, 1983: 36–48.

75 The Brooke family and the British North Borneo Company returned to their respective territories in 1946 but they lacked the resources necessary for reconstruction and establishment of full control. There was some opposition from members of the Brooke family and some Malay officials to the transfer of Sarawak to the British Crown (Andaya and Andaya, 1982: 255).

76 In the immediate post-war period there were six major political groupings in Malaya. The MCP (Malayan Communist Party) and the KMT (Kuomintang) had been established during the 1930s. Three groups were products of the war and the Japanese occupation, MNP (Malay Nationalist Party), MDU (Malayan Democratic Union) and the MIC (Malayan Indian Congress). The most significant organisation was UMNO (United Malay National Organisation). This resistance reflected the rise of nationalist organisations that had developed during the Japanese occupation. The

British view of the 'docile Malays' was rudely shaken by these protests. See Cheah Boon Kheng (1977) for an account of the political organisation active in Malaya between 1945 and 1948.

77 Perhaps more realistically from the 'War of Independence' (see Caldwell and Amin, 1977).

78 'By the end of 1948, it was estimated that the Emergency was costing between M$250,000 and M$300,000 a day. The total cost of defence, police and the Emergency rose from M$82 million in 1948 to a peak of M$296 million in 1953 with the two years 1952 and 1953 being the most expensive. Fortunately for the British authorities in Malaya, much of the revenue for the Emergency expenditure was provided by the Korean war prices boom which lasted for three years – 1950, 1951 and 1952' (Cheah Boon Kheng, 1977: 113).

79 As well as a major destination for British investment the territories were to assume increasing strategic importance following the withdrawal from Burma and India.

80 The initial proposal for the inclusion of Brunei was withdrawn in the face of opposition. In December 1962 there was open rebellion which was subsequently repressed by British forces. In retrospect it is apparent that the failure of Brunei to join the Federation seriously weakened it.

81 For a review of British and American relations with Thailand during this period see Aldrich (1988).

82 Ho had been in exile for thirty years.

83 For an account of the Japanese occupation of Indo-China see McCoy (1981).

84 The control over Saigon was disputed with the French and a variety of nationalist and revolutionary groups.

85 The Vietminh came into open conflict with other nationalist groups and, most significantly, the Trotskyists of the ICL (International Communist League). It is clear that, at this stage, the Vietminh were effectively a 'popular front' with a programme of national liberation and agrarian reform. They thus came violently into conflict with the workers' committee and demands for expropriation of property. There is little doubt that the leadership of the Trotskyists was effectively exterminated by the Vietminh, French and British. (The archive material on these events is not readily accessible. A selection of documents *Vietnam and Trotskyism* was published by the Communist League of Australia in 1987).

86 Under these proposals France would retain jurisdiction over defence, foreign relations and finance.

87 This included US$21.18 million for 'economic aid'.

88 The USA backed off from direct intervention largely because Britain refused to support such a move. However, the fear of a major confrontation with China and the 'winding down' of the Korean war were also of major importance in the decision (Kalb and Abel, 1971: 82).

89 For an excellent account of the war from an international perspective see Smith (1983; 1985).

5 Development strategies and the international economy

1 Despite the preoccupation of the international agencies with Third World levels of protection these are frequently much higher for the developed economies. See for example the 1986 IMF review.

2 This report concluded that in:

> Malaysia, between 1963 and 1971 the tax concessions on pioneer industries cost US$10 million, the equivalent of 36 per cent of the sector's capital.
>
> Thailand, in 1973 tariff exemption for promoted firms amounted to 1.9 per cent of government revenue.
>
> Philippines, in 1975 tax exemption under the Investment Incentive Act cost US$33 million for 126 firms, equivalent to 2.1 per cent of the total tax revenue.

Kraus and Lütkenhorst (1986: 46) suggest that in South Korea the cost of incentives was the equivalent of 22 per cent of export earnings.

3 See McGee (1986 et al.) for a review of Malaysian EPZs.

4 The United Nations' Industrial Survey Mission (1961) *A proposed industrialisation programme for the State of Singapore*, Geneva. This mission was led by Dr Albert Winsemius and in consequence is sometimes referred to as the 'Winsemius Mission'.

5 Between 1959 and 1983 there were no opposition MPs elected.

6 For example, the 1979 Warehouse and Incentive Scheme and the International Consulting Scheme.

7 This was principally public sector led by the Housing Development Board and such major projects as the new airport and the Mass Rapid Transport System.

8 There is little doubt that increasingly the term 'national interest' was a euphemism for 'vested interest' on the part of members of the government, military or well-connected members of the Chinese business community.

9 The 1954 Promotion of Industries Act set out various incentives for both domestic and foreign investors. However, it remained almost totally ineffective until the establishment of the BOI in 1959. Further legislation in 1960 and 1962 widened the scope of the provisions.

10 American Economic Assistance to Thailand increased from US$39.5 million in 1965 to a peak of US$53.8 million in 1967, 67 per cent of which was 'military aid'. These figures do not take account of the expenditure and employment generated by the American presence. In 1967 there were 44,100 American troops in the Kingdom and 43,750 Thais were directly employed by the military. Additionally an enormous volume of service employment was generated.

11 The rundown of the American presence in Thailand accompanied the gradual withdrawal from Vietnam. Serious unemployment resulted in the over-extended service sector. In 1974 an IBRD assistance programme was implemented to rehabilitate the former base towns in the North East.

12 A considerable volume of overseas funds have been channelled into local financial institutions which in turn funded much of the foreign manufacturing industry.

13 'Malays'.
14 Oil palm production grew from 90,000 tons in 1960 to 649,000 in 1972. Since 1966 Malaysia has been the world's largest producer.
15 The long-term target of 80–85 per cent self-sufficiency has been revised to 60–65 per cent. In 1984 Malaysia produced 69 per cent of its requirement and in the 'bumper' harvest of 1985, 80 per cent. These higher levels of self-sufficiency are not now regarded as sustainable.
16 These include PETRONAS (National Oil Corporation), all public utilities and transport organisations, HICOM (Heavy Industries Corporation of Malaysia), Sabah Forest Industries and FIMA (Food Industries of Malaysia).
17 These held ownership 'in trust' for the *bumiputeras* where there was a lack of capital to purchase the allocated shares.
18 Since 1980 anti-labour policies have become even more marked and trade union membership has declined (Sundaram, 1987: 137).
19 See Warr (1986) for a counter-view.
20 The government has been unable to sell the Malaysian railways even at a token M$1.
21 This decline is usually blamed on nationalisation and subsequent mismanagement of estates. The decline in estate production was partly offset by remarkable expansion in the smallholder sector (Palmer, 1978: 11–12).
22 The expulsion of foreign capital from the non-oil sector was not as complete as is sometimes suggested. By the end of the Sukarno period a modest US $72m had been invested, US $30.75m by Japan (Robison, 1986: 79–80).
23 This has been the subject of much research and speculation. See, for example, Mortimer (1974), Brackman (1969), May (1978) and Törnquist (1984).
24

	Percentage share of budget channelled through IGGI
Repelita I	66
Repelita II	35
Repelita III	42
Repelita IV	35

25 In Java unemployment was estimated at 7 per cent and underemployment 30 per cent (Wong, 1979: 59).
26 Data on Indonesian foreign investment are problematic. In chapter 1 comparative data for ASEAN was drawn from the IMF *Balance of Payments Yearbook*; this is almost certainly the most reliable source. However, very different pictures emerge for Indonesia if data from OECD, Bank of Indonesia or BKPM (Capital Investment Coordinating Board) are used. The latter, which because of their ready availability are the most widely cited, are the least accurate. In consequence all Indonesian investment data should be treated with caution. See Hill (1988) for a further discussion.

Estimates of direct foreign investment in Indonesia, 1967–1985 ($ million)

Year	OECD	IMF	Board of Investment	BKPM
1985	−295	271	299	773
1984	423	227	245	1,121
1983	303	289	193	2,882
1982	537	226	311	1,800
1981	2,584	133	142	1,092
1980	300	184	140	914
1979	−383	226	217	1,320
1978	418	279	271	397
1977	−67	235	285	328
1976	746	344	287	449
1975	1,289	476	454	1,757
1974	182	−49	538	1,417
1973	348	15	318	655
1972	90	207	240	522
1971	117	139	173	426
1970	49	83	90	345
1969	48	32	39	682
1968	n/a	−2	−3	230
1967	n/a	−10	−12	125

Source: Hill, 1988: 158

27 The figures used in this paragraph are from the BKPM (Capital Investment Coordinating Board) and should be regarded as approximate.

Gross Foreign Investment, 1968–1984 ($ million)

Year	BKPM sectors 'implemented'	Petroleum	%	Total
1984	388	1,994	83.71	2,382
1983	517	2,728	84.07	3,245
1982	459	2,708	85.51	3,167
1981	379	2,111	84.78	2,490
1980	347	1,435	80.52	1,782
1979	319	745	70.02	1,064
1978	405	593	59.42	998
1977	256	600	70.01	856
1976	426	766	64.26	1,192
1975	547	725	56.97	1,272
1974	634	557	46.77	1,191
1973	394	271	40.75	665
1972	271	197	42.09	468
1971	151	161	51.60	312
1970	98	96	49.48	194

1969	47	54	53.46	101
1968	18	26	59.09	44
Cumulative Total	5,656	15,767		21,423

Source: Hill, 1988: 160. The petroleum data are derived from the United States Embassy annual reports *Indonesia's Petroleum Sector*, Jakarta.

28 The exhaustion of forestry resources in the Philippines resulted in much capital moving into the Indonesian industry.

29 Between 1967 and 1980 64.5 per cent of cumulative non-oil f.d.i. went into manufacturing. Textiles and leather comprised 34.4 per cent of manufacturing f.d.i., metals 18.3 and chemicals and rubber 15.8 (Forbes, 1986: 114).

30 For a summary of these measures see Forbes, 1986: 133–40.

31 The World Bank (1984 and 1985) cited the following subsidies to Indonesian industries

Portland cement	US$156.8 million
Colour TV	US$ 43.6 million
Synthetic yarn	US$ 24.6 million
Steel billets	US$ 74.8 million
Motor cycles	Rp 53.2 billion
Cosmetics	Rp 67.1 billion
Motor vehicles	Rp134.4 billion
Glass	Rp 37.5 billion
Sugar	Rp137.6 billion

32 Repelita V (1989–93) appears to pave the way for greater private investment in the economy and a reduction in state activity. How these policies will be implemented remains unclear (Vatikiotis, 1989).

33 Since 1961 economic policy owed much to continual IMF involvement (Canlas *et al.*, 1988: 23).

34 Until 1963 the import of bananas into Japan was restricted and the market almost entirely supplied from Taiwan (Krinks, 1986: 260).

35 'The first [to be established] was Castle and Cooke, which operated through rather complex arrangements. Its first step was to set up Standard (Philippines) Fruit Corporation (Stanfilco) as a 66 per cent-owned joint venture with a Philippine holding company, House of Investments, in which Castle and Cooke had 17 per cent equity. Its Filipino major partners in House of Investments were already associated with Castle and Cooke through Dole Philippines, Inc. (Dolefil), set up in 1963 with 80 per cent equity held by the American corporation. The Filipinos were lawyers and businessmen associated with the Rizal Commercial and Banking Corporation (RCBC), and included J. P. Enrile, later the Secretary of Justice and the Minister of Defence under President Marcos. The president of Dolefil, G. P. Velasco, later became Marcos's Minister for Energy. In the same year, the government's National Development Corporation acquired 5500 hectares, mostly occupied by small-scale settlers and, almost certainly unconstitutionally, leased the land to Dolefil for 25 years (renewable)' (Krinks, 1986: 261–2).

36 In 1977 83 per cent of the sales of USA-owned plants went to the Filipino market (Jayasuriya, 1987: 95).

37 The new deadline, May 1988, was not met either.

38 Measurement in this area is, of course, a minefield. Totally opposing conclusions can be found for most of the countries in the region; occasionally, and most revealingly, apparently based on the same data.

39 The calculation of poverty lines varies from study to study, making comparisons over time and between countries extremely difficult. In the Philippines, for example, the poverty line has been variously drawn as:
US$77.31 per month for a family of six (Tan and Holazo, 1979)
US$251.04 per month for a family of five (Centre for Research and Communications, 1982 cited Canlas, *et al.*, 1988: 12)
US$198.86 per month for a family of six (World Bank, 1980a).

40 Unfortunately the study does not make it clear whether it is using the World Bank poverty line.

41 The Gini coefficient ranges from 0.0 – absolute equality – to 1.0 – absolute inequality.

42 See Jones (1988: 139) for a discussion of these projections.

6 South East Asia in the late twentieth century: problems and perspectives

1 For example, the relocating of labour intensive shipbuilding and repairing to the Philippines.

2 Information on Johor drawn from *Far Eastern Economic Review* (18 August 1988: 77).

3 It has been estimated that 200 people employed in wafer fabrication can generate as much revenue as 2000 in assembly and testing (Far Eastern Economic Review, 18 August 1988: 85).

4 The prospects for this were greatly enhanced during 1987 with the establishment of the Laotian and Vietnam foreign investment codes. In September 1988 Shell (UK) signed a US$70 million exploration agreement for the Bach Bo oil field (off the Mekong delta).

5 The other ASEAN countries, with the exception of Singapore, have also shown interest in the prospect of new markets, investment locations and raw material sources in these countries. However, Thailand has been by far the most active of the South East Asia countries in this respect and appears even to be prepared to sponsor Vietnam's membership of ASEAN.

6 As well as hydro-electric power produced in Laos.

7 Similar moves are taking place in the printing industry.

References

Abu-Lughod, J. (1988) 'The shape of the world system in the thirteenth century' *Studies in comparative international development* XXII, 4: 3–25.

Adas, M. (1974) *The Burma Delta: economic development and social change on an Asian rice frontier, 1851–1941*, University of Winsconsin Press, Winsconsin.

Aditjondin, G. T. (1984) 'Problems of forestry and land use in the Asia-Pacific region: the Irian Jaya experience' in Sahabet Alam Malaysia (Friends of the Earth, Malaysia) *Environment development and natural resource in crisis in Asia and the Pacific*, Penang, 53–70.

Agnew, J. (1987) *The United States in the world economy*, Cambridge University Press.

Akrasanee, N. and Juanjai, A. (1986) 'Thailand: manufacturing industry protection issues and empirical studies', in C. Findlay and R. Garnaut eds. (1986) *The political economy of manufacturing protection: experiences of ASEAN and Australia*, Allen and Unwin, London: 77–98.

Akrasanee, N. (1980) *Industrial sector in the Thai economy*, Thai University Research Association, Report No 1, Bangkok.

Akashi, Y. (1980) 'The Japanese occupation of Malaya: interruption or transformation', in A. W. McCoy ed. *South East Asia under Japanese occupation*, South East Studies Monograph no. 22, Yale University: 63–85.

Akira, Suehiro (1985) *Capital accumulation and industrial development in Thailand*, Chulalongkorn University Social Research Institute, Bangkok.

Aldrich, R. (1988) 'A question of expedience: Britain, the United States and Thailand, 1941–2', *Journal of South East Asian Studies*, 19, 2: 209–44.

Allen, G. C. and Donithorne, A. G. (1957) *Western enterprise in Indonesia and Malaya*, Macmillan, London.

Amin, S. (1973) *Le Dévéloppement inégal*, Les Editions de Minuit, Paris.

Anand, S. (1983) *Inequality and poverty in Malaysia*, Oxford University Press, London.

Andaya, B. W. and Andaya, L. (1982) *A history of Malaysia*, Macmillan, London.

Andrews, J. M. (n.d.) *Siam: second rural economic survey 1934–35*, Harvard University Press.

Anspach, R. (1969) 'Indonesia', in F. Golay *et al.*, eds. *Economic nationalism in South East Asia*, Cornell University Press, Ithaca: 111–201.

Ariff, M. and Hill, H. (1985) *Export-orientated industrialisation: the ASEAN experience*, Allen and Unwin, London.

ASEAN Business Quarterly, various issues.

Asian Business, various issues.

Aung-Thwin, M. (1985) 'The British "pacification of Burma": order without meaning', *Journal of South East Asian Studies* 16, 2: 245–61.

Ayal, E. B. (1961) 'Some crucial issues in Thailand's economic development', *Pacific Affairs*, 34: 157–64.

Aziz, U. (1962) *Subdivision of estates in Malaya, 1951–60*, 3 volumes, University of Malaya, Kuala Lumpur.

Bailey, A. M. (1981) 'The renewed discussions on the Asiatic Mode of Production', in J. S. Kahn and J. R. Llobera eds. *The Asiatic mode of production*, Routledge and Kegan Paul, London, 89–108.

Bailey, A. M. and Llobera, J. R. eds. (1981a) *The Asiatic mode of production*, Routledge and Kegan Paul, London.

Bailey, A. M. and Llobera, J. R. (1981b) 'The AMP: sources and the formation of the concept', in A. M. Bailey and J. R. Llobera eds. *The Asiatic mode of production*: 13–45, Routledge and Kegan Paul, London.

(1981c) 'The Wittfogel watershed: introduction', in A. M. Bailey and J. R. Llobera eds. *The Asiatic mode of production*: 109–12, Routledge and Kegan Paul, London.

Balakrishnan, N. (1989) 'More work than workers', *Far Eastern Economic Review*, 19 January: 66.

Bangkok Bank Review, various issues.

Bank of Indonesia, *Indonesian financial statistics*, monthly, Jakarta

Bartlett, A. G. *et al.* (1972) *PERTAMINA: Indonesian National Oil*, Amerasia, Jakarta, Singapore and Tulsa.

Bartlett, A. G. *et al.* (1972) *PERTAMINA: Indonesian National Oil*, Amerasia, Jakarta.

Bartalits, L. L. S. and Schneider, J. W. H. C. M. (1987) 'Soviet strategy and the Pacific Basin', in P. West and F. A. M. Alting von Geusau eds. *The Pacific Rim and the Western world*, Westview, London: 43–67.

Bauer, B. T. (1984) 'Is planning the answer?' in *Singapore Business Yearbook 1984*, Times Periodicals, Singapore: 18–24 and 44.

Benda, H. J. and Larkin, J. A. eds. (1967) *The world of South East Asia: selected historical readings*, Harper and Row, London.

Berger, P. L. and Hsiao, H. H. (1988) *In search of an Asian model of development*, Transaction Books, New Brunswick.

Blaut, J. M. (1976) 'Where was capitalism born?' *Antipode*, 8, 2: 1–11.

Board of investment (1966) *Promotion of Investment Act, 1962*, Bangkok.

Boeke, J. O. H. (1944) 'Diversity and unity in South East Asia', *Geographical Review*, 34: 175–95.

(1947) *The evolution of the Netherlands Indies economy*, H. D. Tjeenk Willink, Haarlem.
(1953) *Economics and economic policy of dual societies*, H. D. Tjeenk Willink, Haarlem.
(1966) *Indonesia: the concept of duralism in theory and policy*, Van Hoeve, The Hague.
Bonner, R. (1987) *Waltzing with a dictator: the Marcoses and the making of American policy*, Macmillan, London.
Boomgard, P. (1980) 'Bevolkingsgroei en welvaart op Java (1800–1942) in R. N. J. Kamerling ed. *Indonesie toen en nu* Intermediair, Amsterdam: 45–7.
(1987) 'Morbidity and mortality in Java, 1820–1880: changing patterns of disease and death', in N. G. Owen ed. *Death and disease in South East Asia*, Oxford University Press, Singapore: 48–59.
Booth, A. (1980) 'The burden of taxation in Colonial Indonesia', *Journal of South East Asian Studies*, 11, 1: 91–8.
(1988) 'Living standards and the distribution of income in Colonial Indonesia: a review of the evidence', *Journal of South East Asian Studies*, 19, 2: 310–34.
Borromeo-Buehler, S. (1985) 'The Inquitinos of Cavite: a social class in nineteenth-century Philippines', *Journal of South East Asian Studies*, 16, 1: 69–98.
Bowring, P. (1982) 'ASEAN's dark horse', *Far Eastern Economic Review*, 85, 22 October.
Boxer, C. R. (1969) *The Portuguese seaborne empire, 1415–1825*, Knopf, New York.
Brackman, A. C. (1969) *The Communist collapse in Indonesia*, Norton, New York.
Bradley, P. N. (1986) 'Food production and distribution – and hunger', in R. J. Johnston and P. J. Taylor eds. *A world in crisis*, Blackwell, Oxford: 89–106.
Braudel, F. (1979) *Civilization and capitalism, 15th–18th century*, in three volumes, English translation 1982, 1983 and 1984, Collins, London.
Bray, F. (1986) *The rice economies: technology and development in Asian societies*, Blackwell, Oxford.
Brennan, M. (1985) 'Class, politics and race in modern Malaysia', in R. Higgott and R. Robinson eds. *South East Asia: essays in the political economy of structural change*, Routledge and Kegan Paul, London: 93–129.
Brenner, R. (1977) 'The origins of capitalist development: a critique of neo-Smithian Marxism', *New Left Review* 104: 25–93.
Brown, I. (1978) 'British financial advisers in Siam in the reign of King Chulalongkorn', *Modern Asian Studies*, 12, 2: 193–215.
(1986) 'Rural distress in South East Asia during the world depression of the early 1930s: a preliminary re-examination', *Journal of Asian Studies*, 45, 5: 995–1025.
(1988) *The elite and the economy of Siam c.1890–1920*, Oxford University Press.

(1989) *The economics of Africa and Asia in the inter-war depression*, Routledge, London.

Buchanan, K. (1967) *The South East Asian world*, G. Bell and Sons, London.

Buchanan, I. (1972) *Singapore in South East Asia*, G. Bell and Sons, London.

(1987) 'Indonesia: economy', in *Far East and Australia 1987*, Europa, London: 441–54.

Bunout, R. (1936) *La main-d'oeuvre et la législation du travail en Indochine*, Imprimerie-Librairie Delmas, Bordeaux.

Buss. C A. ed. (1985) *National security interests in the Pacific Basin*, Hoover Institute Press, Stanford, California.

Buttinger, J. (1958) *The smaller dragon: a political history of Vietnam*, Prager, New York.

(1967) *Vietnam: a dragon embattled*, Praeger, New York.

Cady, J. F. (1958) *A history of modern Burma*, University of California Press, Ithaca.

Caldwell, J. A. (1974) *American economic aid to Thailand*, Lexington Books, Lexington, Massachusetts.

Caldwell, J. C. (1963) 'Urban growth in Malaya: trends and implications' *Population Review*, 17: 39–50.

Caldwell, M. (1978) 'Foreword', in D. Elliott, *Thailand: origins of military rule*, Zed, London: 8–20.

Caldwell, M. and Amin, M. (1977) *Malaya: the making of a non-colony*, Spokesman, London.

Callis, H. (1942) *Foreign capital in South East Asia* International Secretariat, Institute of Pacific Relations, New York.

Canlas, M. Mirand, M. and Putzel, J. (1988) *Land, poverty and politics in the Philippines*, Catholic Institute for International Relations, London.

Carey, P. B. R. (1979) 'Aspects of Javanese history in the 19th century' in H. Aveling ed. *The development of Indonesian society*, University of Queensland Press, St Lucia: 45–105.

Cass, A. (1985) 'Singapore: economic challenges ahead' *Financial Times* 6 November.

Chai Hon Chan (1967) *The development of British Malaya, 1896–1909*, Oxford University Press, Kuala Lumpur.

Chandra, B. (1981) 'Karl Marx, his theories of Asian societies and colonial rule', *Review*, 5 1: 13–91.

Chandra, N. (1986) 'Bumiputra policies versus industrial development: high cost of equity', *Far Eastern Economic Review*, 133, 32: 44–5.

Chatthip Nartsupha and Suthy Prasartset (1981) *The political economy of Siam, 1851–1910*, Social Sciences Association of Thailand, Bangkok.

Cheah Boon Kheng (1988) 'The erosion of ideological hegemony and Royal power and the rise of Malay Nationalism', *Journal of South East Asian Studies*, 14, 1: 1–26.

(1977) 'Political parties in Malaya: 1945–1948', in Kay Kim Khoo ed. *The history of South East, South and East Asia*, Oxford University Press, Kuala Lumpur: 95–110.

(1977) 'Some aspects of the Malayan emergency: 1948–1949', in Khoo Kay Kim ed. *The history of South East and South Asia*, Oxford University Press, Kuala Lumpur: 111–21.

Chee Khoon and Chan Chee Heng Leng (1984) *Designer genes: IQ, ideology and biology*, Institute for Social Analysis, Malaysia.

Chesneaux, J. (1966) *The Vietnamese nation: a contribution to history*, Current Books, Sydney.

Chew, E. C. T. (1979) 'The fall of the Burmese Kingdom in 1885: review and reconstruction', *Journal of South East Asian Studies*, 10 2: 372–80.

Chia Siow-Yue (1986) 'EC direct private investment in ASEAN states: a South East Asian view', in R. H. Taylor and P. C. I. Ayre eds. *ASEAN-EC economic and political relations*, School of Oriental and African Studies, London: 91–110.

Chin Yong Fong, (1977) 'Dutch policy towards the Outer Islands', in Khoo Kay Kim ed. *The history of South East and South Asia*, Oxford University Press, Kuala Lumpur: 129–35.

Cho, G. (1990) *The Malaysian economy: spatial perspectives*, Routledge, London.

Cipolla, C. M. (1965) *Guns and sails in the early phase of European expansion, 1400–1700*, Collins, London.

Collier, W. L. (1975) 'Tebasan system, high yielding varieties and rural change', *Prisma*, 1.

1978) 'Food problems, unemployment and the Green revolution in Java', *Prisma*, 9.

Communist League of Australia (1987) *Vietnam and Trotskyism*, Petersham, New South Wales.

Constantino, R. and Constantino, L. R. (1978) *The Philippines: the continuing past*, Foundation for Nationalist Studies, Manila.

Council for Economic Planning and Development, Republic of China (1985) *Taiwan statistical data book*, Tapei.

Courtenay, P. P. (19723) *A geography of trade and development in Malaya*, Bell, London.

(1979) 'Some trends in the peninsular Malaysia plantation sector', in J. C. Jackson and M. Rudner eds. *Issues in Malaysian development*, Heinemann, Singapore: 131–65.

Cowan, C. D. (1961) *Nineteenth-century Malaya: the origins of British political control*, Oxford University Press, London.

Crawfurd, J. (1828) *History of the Indian Archipelago*, Archibald Constable and Co., Edinburgh, 3 volumes.

Creutzberg, P. ed. (1975) *Changing economy of Indonesia: a selection of statistical source material from the early 19th century up to 1940, Volume I, Indonesia's export crop 1816–1940*, Martinus Nighoff, The Hague.

Cunningham, S. (1983) 'Technical change and Newly Industrialising Countries: the division of labour and the growth of indigenous technological capability', paper presented at the International Geographical Union – Association of American Geographers' Conference held at Denver, Colorado.

Cushner, N. P. (1971) *Spain in the Philippines: from conquest to revolution*, Institute of Philippine Culture, Areneo de Manila University.

Davenport-Hines, R. P. T. and Jones, G. eds. (1989) *British Business in Asia since 1860*, Cambridge University Press.

Delvert, D. (1961) *Le Paysen cambodgien*, Mouton, Paris.

De Rosta, D. A. (1986) 'Trade and protection in Asian Development Region' *Asian Development Review*, 4, 1.

Devan, J. (1987) 'The economic development of the ASEAN countries', in P. West and F. A. M. Alting von Geusau eds. *The Pacific Rim and the Western world*, Westview, London: 159–78.

Dhoquios, G. (1970) *Deux ouvrages de référence sur les societés pré-capitalistes, La Pensée*, 154: 110–18.

Dixon, C. J. (1977) 'Development, regional disparity and planning: the experience of North East Thailand' *Journal of South East Asian Studies*, 8, 2: 210–23.

(1981) 'Capitalist penetration, uneven development and government response: the case of Thailand', in M. B. Gleave ed. *Societies in change: studies of capitalist penetration*, Developing Areas Research Group Monograph No 2: 65–92.

(1984) 'The Far East after the boom years', *Geographical Magazine*, 52: 61–6.

(1990a) *Rural development in the Third World*, Routledge, London.

(1990b) 'Human resources', in D. J. Dwyer, ed. *South East Asian development*, Longman, London: 149–85.

Donges, J. B., Stecher, B. and Wolter, F. (1980) 'Industrialisation in Indonesia., in G. Papanek ed. *The Indonesian Economy*, Praeger, New York: 357–405.

Donner, W. (1973) *The five faces of Thailand: an economic geography*, G. Hurst, London.

Drabble, J. H. and Drake, P. J. (1981) 'The British Agency Houses in Malaysia: survival in a changing world', *Journal of South East Asian Studies*, 12, 2: 297–328.

Drakakis-Smith D. W. and Rimmer, P. J. (1982) 'Taming "The wild city": managing South East Asia's primate cities', *Asian Geographer*, 1, 1: 17–34.

Drakakis-Smith, D. W. (1990) 'Urban essays for Harold Carter' in C. R. Lewis ed. *Colonial urbanisation in Asia and Africa: a structural analysis*, Cambria, University of Wales.

Drake, P. J. (1979) 'The economic development of British Malaya', *Journal of South East Asian Studies*, 10, 2: 262–90.

Dumont, R. (1935) *La Culture de riz dans de Delta du Tonkin*, Société d'Editions Géographiques, Maritimes and Coloniales, Paris.

Dunn, C. W. (1928) 'Some general characteristics of agricultural economies in Burma' in *Agriculture in Burma: a collection of papers written by government officials for the Royal Commission on Agriculture 1926–28*, Government of Burma, Rangoon: 15–27.

Eberhard, W. (1958) 'Oriental despotism: political weapon or sociological concept?', *American Sociological Review*, 23: 446–8.

Economist (1983) 'A question of breeding', 24 September: 50.

Elliott, C. B. (1963) *The Philippines: a study in tropical democracy*, Greenwood Press, New York. First published in 1917 by Boobs-Merrill, Inc.

Elliott, D. (1978) *Thailand: origins of military rule*, Zed, London.

Emmerson, D. K. (1984) '"South East Asia": what's in a name?' *Journal of South East Asian Studies*, 15, 1: 1–21.

ESCAP (1983) *Foreign investment incentive schemes*, United Nations, Bangkok.

(1987) *International labour migration and remittances between the developing ESCAP countries and the Middle East: trends, issues and policies*, United Nations, Bangkok.

(1988) *Urban geology in Asia and the Pacific: atlas of urban geology*, United Nations, Bangkok.

Evers, Hans-Dieter (1987) 'Trade and state formation in the early Bangkok period', *Modern Asian Studies*, 21, 4: 751–71.

Falkus, M. (1989) 'Early British business in Thailand', in R. P. T. Davenport-Hines and G. Jones eds. *British Business in Thailand*, Cambridge University Press: 117–55.

(1989) 'Early business in Thailand', in R. P. T. Davenport-Hines and G. Jones *British business in Asia since 1860*, Cambridge University Press: 92–116.

FAO *Production Year Book*, Rome, various issues.

Far Eastern Economic Review Year Book, Hongkong, various issues.

Farris, W. W. (1985) *Population, disease and land in early Japan*, Harvard University Press, Cambridge.

Feeney, D. (1982) *The political economy of productivity: Thai agricultural development, 1880–1975*, University of British Columbia Press, London and Vancouver.

(1979) 'Paddy prices and productivity: irrigation and Thai agricultural development', *Explorations in economic history* 16: 132–50.

Far Eastern Economic Review (1983) 'A peek into the grey zones', 12 May: 59–60.

Fernando, R. (1988) 'Rice cultivation in Cirebon Residency, West Java, 1830–1940', *RIMA* 22, 1: 1–68.

Financial Times (1981), 29 June: 11.

Findlay, C. and Garnaut, R. eds. (1986) *The political economy of manufacturing protection: experiences of ASEAN and Australia*, Allen and Unwin, London.

Fisher, C. A. (1962) 'South East Asia: the Balkans of the Orient', *Geography* 47: 347–67.

(1964) *South East Asia: a social, economic and political geography*, Methuen, London.

(1968) 'Malaysia: a study in the political geography of decolonisation', in C. A. Fisher ed. *Essays in political geography*, Methuen, London: 75–146.

(1971) 'South East Asia', in W. G. East, O. H. K. Spate and C. A. Fisher eds. *The changing map of Asia*, Methuen, London: 221–337.

Fitzgerald, C. P. (1972) *The southern expansion of the Chinese people*, Barrie and Jenkins, London.

Forbes, D. (1986) 'Spatial aspects of Third World multinational corporations' direct investment in Indonesia', in M. Taylor and N. Thrift *Multinationals and the restructuring of the world economy*, Croom Helm, London: 86–104.

Forbes, D., Kissling, C. C., Taylor, M. J. and Thrift, N. J. (1985) *Economic and social atlas of the Pacific Basin*, Allen and Unwin, Sydney.

Foster, G. M. (1965) 'Peasant society and the image of the limited good', *American Anthropologist* 67: 293–315.

Frankfurter, O. (1904) 'King Mongkut', *Journal of the Siam Society*, 1: 191–208.

Fryer, D. W. (1979) *Emerging South East Asia*, Wiley, London.

Furnivall, J. S. (1936) 'The weaving and batik industries in Java' *Asiatic Review*, 33: 373–4.

(1938) *An introduction to the political economy of Burma*, Burma Book Club, Rangoon.

(1944) *Netherlands India: a study of plural economy*, Cambridge University Press.

(1948) *Colonial policy and practice: a comparative study of Burma and the Netherlands Indies*, Cambridge University Press.

(1956) *Colonial policy and practice: a comparative study of Burma and Netherlands India*, New York University Press.

Galang, J. (1988) 'The generation gap: growing power shortages threaten Philippines recovery', *Far Eastern Economic Review*, 3 November: 106.

Gale, B. (1981) 'PETRONAS: Malaysia's national oil' *Asian Survey*, 21 11: 1129–44.

Gallagher, J. and Robinson, R. (1953) 'The Imperialism of free trade', *Economic History Review*, 6 1: 1–15.

Garnaut, R. ed. (1980) *ASEAN in a changing Pacific and world economy* Australian National University, Canberra.

GATT (1988) *Review of developments in the trading system*, Geneva.

Geertz, C. (1963) *Agricultural involution: the process of ecological change in Indonesia*, University of California Press, Berkeley.

Gerbrandy, P. S. (1950) *Indonesia*, Hutchinson, London.

Gettleman, M. E., Franklin, J., Young, M. and Franklin, H. B. (1985) *Vietnam and America: a documented history*, Grove Press, New York.

Gibbons, D., De Koninck, R. and Hasan, I. (1986) *Agricultural modernization, poverty and inequality*, Gower, London.

Gill, R. (1985) *The making of Malaysia Inc.*, Pelanduk Publications, Singapore.

Ginsberg, N. S. (1955) 'The great city in South East Asia', *American Journal of Sociology*, 60 5: 455–62.

Glamann, K. (1971) *European trade, 1500–1700*, Fontana, London.

Glassburner, B. (1971) 'Indonesian economic policy after Sukarno', in B. Glassburner ed. *The economy of Indonesia: selected readings*, Cornell University Press, Ithaca: 425–43.

Godelier, M. (1977) *Perspectives in Marxist anthropology*, Cambridge University Press.

(1978) 'The concept of the Asiatic Mode of Production', in D. Seddon ed. *The relations of production*, Cass, London.

Godinho, V. M. (1969) *L'Économie de l'empire portugais aux XVe et XVIe siècles*, SEVPEN, Paris.

Goh Keng Swee (n.d.) 'Some problems of industrialization' (the text of six radio talks by the Minister for Finance, 2–12 January 1963), Ministry of Culture, Singapore.

Golay, F. H. (1961) *The Philippines: public policy and national economic development*, Cornell University Press, Ithaca.

Golay, F. H., Anspach, R., Pfannes, M. R. and Ayal, E. B. (1969) *Underdevelopment and economic nationalism in South East Asia*, Cornell University Press, Ithaca.

Goldman, M. F. (1972) 'Franco-British rivalry over Siam, 1896–1904' *Journal of South East Asian Studies*, 3 2: 210–28.

Gorman, C. (1977) '*A priori* models and Thai prehistory: beginnings of agriculture', in C. Reed ed. *Origins of agriculture*, Mouton, The Hague: 321–55.

Gould, J. S. (1950) Estimates of Gross National Product and Net National Income of Thailand 1938–9, 1946, 1947 and 1948, National Economic Council, Bangkok.

Gourou, P. (1936) *Les Paysans du Delta Tonkinois*, Publications de l'Ecole Française d'Extrême-Orient, Paris.

(1958) *The tropical world*, second edition, Longman, London, first published in 1953.

Grahame, W. A. (1924) *Siam*, 2 volumes, Alexander Moring, London.

Griffith-Jones, S. (1984) 'Debt-led growth', paper presented at the Development Studies Association Annual Conference, September 1984, University of Sussex.

Grigg, D. B. (1974) *The agricultural systems of the world*, Cambridge University Press.

Groslier, B. P. (1979) 'La Cité hydraulique angkorienne: exploitation or sure exploitation du sol', *Bulletin de l'Ecole Française d'Extrême-Orient*, 66: 161–202.

(1974) 'Agriculture et religion dans l'Empire angkorien', *Etudes rurales*, 53–6, *Agriculture et sociétés en Asia Sud-est*: 95–117.

Grunder, G. A. and Livezey, W. E. (1951) *The Philippines and the United States*, University of Oklahoma Press.

Guillaumat, P. (1938) 'L'Industrie minérale de l'Indochine en 1937', *Bulletin économique de l'Indochine* 41: 1250.

Hainsworth, G. B. (1979) 'Economic growth and poverty in South East Asia: Malaysia, Indonesia and the Philippines', *Pacific Affairs*, 52 1: 5–41.

Hall, D. G. F. (1981) *A history of South East Asia*, Macmillan, London.

Hall, K. R. (1982) 'The Indianization of Funan: an economic history of South East Asia's first state', *Journal of South East Asian Studies* 13 1: 81–106.

Halliday, J. and McCormack, G. (1973) *Japanese Imperialism today*, Penguin Books, Harmondsworth.

Hamilton, C. (1983a) 'Capitalist industrial development in the Four Little Tigers of East Asia', *Journal of South East Asian Studies*, 13 1: 35–73.

(1983b) 'Capitalist industrial development in the Four Little Tigers of East

Asia', in P. Limqueco and B. McFarlane, *Neo-Marxist theories of underdevelopment*, Croom Helm, London: 137–80.

Handley, P. (1988) 'Overburdened infrastructure threatens to curb rapid economic growth' *Far Eastern Economic Review* 141 39: 94–5.

Hanks, L. M. (1972) *Rice and man: agricultural ecology in South East Asia*, Aldine-Atherton, Chicago.

Hardjono, J. M. (1977) *Transmigration in Indonesia*, Oxford University Press, Kuala Lumpur.

(1983) 'Rural development in Indonesia: the 'top-down' approach', in D. S. M. Lea and D. P. Chaudhri eds. *Rural development and the state*, Methuen, London: 38–65.

Harris, N. (1986) *The end of the third world: newly industrializing countries and the decline of an ideology*, Penguin, Harmondsworth, Middlesex.

Harrison, B. (1963) *South East Asia: a short history*, Macmillan, London.

Harriss, J. (1982) *Rural development: theories of peasant economy and agrarian change*, Hutchinson, London.

Henderson, J. W. (1986) 'The new international division of labour and American semiconductor production in South East Asia', in C. J. Dixon, D. Drakakis-Smith and H. D. Watts, *Multinational Corporations and the Third World*, Croom Helm, London: 91–117.

(1989) *The globalisation of high technology production: society, space and semiconductors in the restructuring of the modern world*, Routledge, London.

Hennart, J.-F. (1986) 'Internationalization in practice: early direct foreign investments in Malaysian Tin Mining', *Journal of International Business Studies*, 17 2: 131–43.

Hewison, K. J. (1985) 'The state and capitalist development in Thailand', in R. Higgott and R. Robinson eds. *South East Asia: essays in the political economy of structural change*, Routledge and Kegan Paul, London: 266–94.

(1987) 'National interests and economic downturn: Thailand', 52–79 in R. Robison, K. Hewison and R. Higgott eds. *South East Asia in the 1980s: the politics of economic crisis*, Allen and Unwin, London: 52–79.

Higgott, R. and Robison, R. (1985) 'Introduction', in *idem*, eds. *South East Asia: essays in the political economy of structural change*, Routledge and Kegan Paul, London: 3–15.

Higham, C. F. W. (1984) 'Prehistoric rice cultivation in South East Asia', *Scientific American* 250 4: 100–7.

Hill, H. (1984) 'Industrialisation in Burma in historical perspective', *Journal of South East Asian Studies* 15 1: 134–49.

(1988) *Foreign Investment and industrialisation in Indonesia*, Oxford University Press, Singapore.

Hill, H. and Johns, B. (1985) 'The role of direct foreign investment in developing East Asian countries', *Weltwirtschaftliches Archiv* 12 2: 355–81.

Hill, J. A. P. (1986) 'Singapore as a premier financial services centre', paper presented at the Singapore Futures Conference, Commonwealth Institute, London, 30–1 January 1986.

Hill, R. D. (1977) *Rice in Malaya: a study in historical geography*, Oxford University Press, Kuala Lumpur.

Hindness, B. and Hirst, P. (1975) *Pre-capitalist modes of production*, Routledge and Kegan Paul, London.

Hoffman, L. and Tan, S. E. (1980) *Review of industrial growth, employment and foreign investment in Peninsular Malaysia*, Oxford University Press, Kuala Lumpur.

Holm, D. F. (1977) *The role of the state railways in Thai history 1892–1932*, unpublished Ph.D thesis, Yale University.

Hong Lysa (1984) *Thailand in the nineteenth century*, Institute of South East Asian Studies, Singapore.

IBON Databank Phil. Inc. (1984) 'Poverty threshold', *Facts and figures*, 153.

IMF (1986) *Exchange arrangements and exchange restorations*, Washington.

Balance of payments yearbook, Washington, various issues.

Directory of trade statistics year book, Washington, various issues.

International finance statistics year book, Washington, various issues.

Industrial Estate Authority of Thailand (n.d.) *Industrial Estate Authority of Thailand*, Bangkok.

Ingleson, J. (1988) 'Urban Java during the depression', *Journal of South East Asian Studies*, 19 2: 292–309.

Ingram, J. C. (1971) *Economic change in Thailand 1850–70*, Stanford University Press, Stanford.

Ishii, Y. (1978) *Thailand: a rice growing society*, University of Hawaii Press, Honolulu.

Jackson, J. C. (1968) *Planters and speculators*, University of Malaya Press, Kuala Lumpur.

(1970) *Chinese in the West Borneo Goldfields*, University of Hull, Occasional Papers in Geography No. 15.

Jacoby, E. H. (1961) *Agrarian unrest in South East Asia*, Asia Publishing House, Bombay.

Jayasuriya, S. K. (1987) 'The politics of economic policy in the Philippines during the Marcos era', in R. Robison, K. Hewison and R. Higgott eds. *South East Asia in the 1980s: the politics of economic crisis*, Allen and Unwin, London: 80–112.

Jeffrey, R. ed. (1981) *Asia: the winning of independence*, Macmillan, London.

Jenkins, R. (1984) 'Divisions over the International Division of Labour', *Capital and Class*, 22: 28–57.

Jeshurun, C. (1985) 'The interests and policies of Malaysia: a study in historical change', in C. A. Buss ed. *National security interests in the Pacific*, Hoover Institution Press, Stanford, California: 197–206.

Johnson, D. B. (1975) 'Rural society and rice economy in Thailand', unpublished Ph.D. dissertation, Yale.

(1981) 'Rice cultivation in Thailand', *Modern Asian Studies*, 15 1: 107–26.

Jones, G. W. (1988) 'Urbanization trends in South East Asia: some issues for policy', *Journal of South East Asian Studies*, 19 1: 137–54.

Journal of South East Asian Studies (1973) 'Japan and the Western Powers in South East Asia', special issue, 9 2.

Journal of South East Asian Studies (1984) 'Symposium on societal organisation in Mainland South East Asia prior to the eighteenth century', 15 2: 219–330.

Kahin, G. McT (1952) *Nationalism and revolution in Indonesia*, Cornell University Press, Ithaca.

(1986) *Intervention: how America became involved in Vietnam*, Alfred A. Knopf, New York.

Kahn, J. S. (1981) 'Mercantilism and the emergence of servile labour in colonial Indonesia', in J. S. Kahn and J. R. Llobera eds. *The anthropology of pre-capitalist societies*, Macmillan, London: 185–213.

Kahn, J. S. and Llobera, J. R. eds. (1981) *The anthropology of pre-capitalist societies*, Macmillan, London.

Kalb, M. and Abel, E. (1971) *Roots of involvement: the US in Asia, 1784–1971*, Pall Mall, London.

Karnov, S. (1983) *Vietnam: a history*, Guild Publishing, London.

Kerkvliet, B. J. (1977) *The Huk Rebellion: a study of peasant revolt in the Philippines*, California University Press, Berkeley.

Kaul, M. M. (1982) 'Indians in South East Asia: the colonial period and its impact', in I. J. B. Singh ed. *Indians in South East Asia*, Sterling Publishers, New Delhi: 28–33.

Keyes, C. F. (1965) *Isan: regionalism in North East Thailand*, Cornell University Press, Ithaca.

Khoi, Le Thanh (1973) 'A contribution to the study of the AMP: the case of ancient Vietnam', in A. M. Bailey and J. R. Llobera eds. *The Asiatic mode of production*, Routledge and Kegan Paul, London: 281–9. Originally published in French in *La Pensée* 171, 1973: 128–9, 133–4, 135 and 136–40.

Khoo Kay Kim (1972) '*The Western Malay States 1850–1873: the efforts of communal development on the Malay politica*, Oxford University Press, Kuala Lumpur.

(1977a) 'Descriptive account of the Eastern Malay States', in Khoo Kay Kim ed. *The history of South East and South Asia*, Oxford University Press, Kuala Lumpur: 22–40.

(1977b) 'The Great Depression: the Malaysian context', in Khoo Kay Kim ed. *The history of South East and South Asia*' Oxford University Press, Kuala Lumpur: 78–94.

ed. (1977c) *The history of South East and South and East Asia*, Oxford University Press, Kuala Lumpur.

Kiernan, V. G. (1974) *Marxism and imperialism*, St Martin's Press, New York.

Kiljunen, K. (1984) *Kampuchea: decade of the genocide*, Zed, London.

Kirk, W. (1990) 'South East Asia in the colonial period: a study of cores and peripheries in the development process' in D. W. Dwyer ed. *South East Asia*, Longman, London, in press.

Krader, L. (1975) *The Asiatic Mode of Production*, Van Gorkum, Assen.

(1981) 'Principles and critique of the Asiatic mode of production', in A. M. Bailey and J. R. Llobera eds., *The Asiatic mode of production*, Routledge and Kegan Paul, London: 325–34.

Krataska, P. H. (1983) 'Ends that we cannot foresee: Malay Reservations in British Malaya', *Journal of South East Asian Studies*, 14, 1: 149–68.

Kraus, W. and Lütkenhorst, W. (1986) *The economic development of the Pacific Basin: growth dynamics, trade relations and emerging cooperation*, C. Hurst and Company, London.

Krinks, P. (1983) 'Rectifying inequality or favouring the few? Image and reality in Philippine development', in A. M. Lea and D. P. Chaudhri *Rural development and the state*, Methuen, London: 100–26.

(1986) 'Fruits of independence? Philippine Capitalists and the banana export industry', in M. Taylor and N. Thrift eds. *Multinationals and the restructuring of the world economy*, Croom Helm, London: 256–81.

Kurien, J. (1984) 'Water, fisheries and the depletion of marine resources', in Sahabat Alam Malaysia *Environment, development and natural resource crisis in Asia and the Pacific*, Penang: 95–106.

Lafeber, W. (1963) *The new empire: an interpretation of American expansion 1860–1989*, Cornell University Press, Ithaca.

Lal, A. H. (1975) 'Problems of federal finance in the Malaysian plural society', in D. Lim ed. *Reading on Malaysian economy*, Oxford University Press, Kuala Lumpur: 399–407.

Landon, K. P. (1939) *Siam in transition*, University of Chicago Press, Chicago.

Leach, E. (1959). 'Hydraulic society in Ceylon', *Past and Present*, 15: 16–25.

Lee Kiong Hock (1986) 'The structure and causes of manufacturing sector production', in C. Findlay and R. Garnaut eds. *The political economy of manufacturing protection: experiences of ASEAN and Australia*, Allen and Unwin, London: 99–134.

Lee Oey Hong (1981) *War and diplomacy in Indonesia: 1945–50*, Asian Monograph Series Number 10, James Cook University, Townsville.

Lee, Y. L. (1979) 'Offshore boundary dispute in South East Asia' *Journal of South East Asian Studies*, 10 1: 175–89.

Legge, J. D. (1972) *Sukarno: a political biography*, Penguin Books, Harmondsworth.

Lehman, J-P. (1986) 'Problems in ASEAN-EC relations', in R. H. Taylor and P. C. I. Ayre eds. *ASEAN-EC economic and political relations*, School of Oriental and African Studies, London: 47–62.

Leifer, M. (1989) 'Indonesia rules the waves', *Far Eastern Economic Review*, 5 January: 17.

Leroy-Beaulieu, P. (1891) *De le colonisation chez les peuples modernes*, Guillaumin, Paris.

Lim Chin Choo (1979) 'A comparative analysis of the Shah Alam and Senawang industrial estates', unpublished M.Ec. thesis, University of Malaysia.

Lim Chong Yah (1965) *Economic development of modern Malaysia*, Oxford University Press, Kuala Lumpur.

Lim, D. (1973) *Economic growth and development in West Malaysia*, Oxford University Press, Kuala Lumpur.

Lim, H. K. (1978) *The evolution of the urban system in Malaya*, Penebit University of Malaya, Kuala Lumpur.

Lim, R. (1985) 'The debate on the political economy of Australia-ASEAN relations', in R. A. Higgott and R. Robison eds. *South East Asia: essays in the political economy of structural change*, Routledge, and Kegan Paul, London: 226–38.
Lindblad, J. T. (1989) 'Economic aspects of Dutch expansion in Indonesia, 1870–1914', *Modern Asian Studies*, 23, 1: 1–23.
Lindsey, C. W. (1983) 'In search of dynamism: foreign investment in the Philippines' *Pacific Affairs*, 56: 477–94.
Lipton, M. (1968) 'The theory of the optimizing peasant', *Journal of Development Studies*, 4: 327–51.
Low, L. (1987) 'Malaysia: economy', in *Far East and Australia 1987*, Europa, London: 621–7.
Luxemburg, R. (1913) *The accumulation of capital*, Routledge and Kegan Paul, London (1951).
Macaulay, R. H. (1934) *History of the Bombay Burmah Trading Corporation Ltd 1864–1910*, London.
Mackrell, K. (1986) 'Private sector investment in ASEAN energy development', in R. H. Taylor and P. C. I. Ayre eds. *ASEAN-EC economic and political relations*, School of Oriental and African Studies, London: 137–46.
Majid, S. and Majid, A. (1983) 'Public sector land settlement: rural development in West Malaysia', in D. A. M. Lea and D. P. Chaudri eds. *Rural development and the state*, Methuen, London: 66–99.
Malaysia, (1976) *Third Malaysia Plan, 1976–80*, Government Printers, Kuala Lumpur.
Malaysia, (1981) *Fourth Malaysia Plan, 1981–85*, Government Printers, Kuala Lumpur.
Malaysia, (1986) *Fifth Malaysia Plan, 1986–90*, Government Printers, Kuala Lumpur.
Malou, X. (1986) 'EC direct private investment in ASEAN states: a European view', in R. H. Taylor and P. C. I. Ayre eds. *ASEAN–EC economic and political relations*, School of Oriental and African Studies, London: 111–36.
Mandel, E. (1971) 'The AMP and the historical pre-conditions for the rise of capital', in *The formation of the economic thought of Karl Marx*, New Left Books, London: 116–39.
Marr, D. (1971) *Vietnamese anti-colonialism: 1885–1925*, University of California Press, Berkeley.
(1981) 'Vietnam: harnessing the whirlwind', in R. Jeffrey ed. *Asia: the winning of independence*, Macmillan, London: 163–206.
Martin, L. G. (1987) *The ASEAN success story: social, economic and political dimensions*, University of Hawaii Press.
Marzouk, G. A. (1972) *Economic development and policies: case study of Thailand*, Rotterdam University Press, Rotterdam.
Marx, K. (1859) *A contribution to the critique of political economy*, Lawrence and Wishart, London, 1977.
Maurand, P. (1938) 'L'Indochine forestière', *Bulletin économique de l'Indochine*, 41.

May, B. (1978) *The Indonesian tragedy*, Graham Bush, Singapore.

McCawley, P. (1978) 'Some consequences of the Pertamina Crisis in Indonesia', *Journal of South East Asian Studies* 9, 1: 1–27.

McCoy, A. W. (1981) 'The Philippines: independence without decolonisation', in R. Jeffrey ed. *Asia: the winning of independence*, Macmillan, London: 23–62.

McCoy, A. W. ed. (1980) *South East Asia under Japanese occupation*, South East Asian Studies Monograph No.22, Yale University.

McCue, A. (1978) 'Shortage of workers pinches development of Singapore', *Asian Wall Street Journal*, 31 October.

McGee, T. G. (1967) *South East Asian City*, Bell, London.

(1986) 'Joining the global assembly line: Malaysia's role in the international semi-conductor industry', in T. G. McGee *et al. Industrialization and labour force processes: a case study of Peninsular Malaysia*, Research Papers on Development in East Java and West Malaysia, No.1 Australian National University, Canberra: 35–70.

McGee, T. G. *et al.* (1986) *Industrialization and labour force processes: a case study of Peninsular Malaysia*, Research Papers on Development in East Java and West Malaysia, No.1, Australian National University, Canberra.

McMahon, R. J. (1981) *Colonialism and Cold War: the United States and the struggle for Indonesian independence, 1945–49*, Cornell University Press, Ithaca.

Means, G. P. ed. (1978) *The past in South East Asia's present*, Canadian Council for South East Asian Studies, Ottawa.

Meier, G. M. (1976) *Leading issues in development*, Oxford University Press, New York.

Meilink-Roelofsz, M. A. P. (1962) *Asian trade and European influence in the Indonesian Archipelago between 1500 and about 1630*, Martinus Nijhoff, The Hague.

Melotti, U. (1977) *Marx and the Third World*, Macmillan, London.

Mende, T. (1955) *South East Asia Between Two Worlds*, Turnstile Press, London.

Mirza, H. (1986) *Multinationals and the growth of the Singapore economy*, Croom Helm, London.

Mitchell, K. (1942) *Industrialisation of the Western Pacific*, International Secretariat, Institute of Pacific Relations, New York.

Moffat, A. L. (1961) *Mongkut, King of Siam*, Cornell University Press, Ithaca.

Montes, M. F. (1985) 'Rural productivity-raising programme in the Philippines', in S. Mukhopadhyay ed. *The poor in Asia: productivity raising programmes and strategies*, Asia and Pacific Development Centre, Kuala Lumpur: 625–93.

Morello, T. (1983) 'Sweatshops in the sun', *Far Eastern Economic Review* 15 September: 88–9.

Mortimer, R. (1974) *Indonesian Communism under Sukaine. Ideology and politics, 1959–65*, Cornell University Press, Ithaca.

Mougne, C. (1982) 'The social and economic correlates of demographic change in a Northern Thai community', unpublished Ph.D thesis, School of Oriental and African Studies, University of London.

Mukhopadyay, S. (1985) 'An overview of productivity raising programmes for the poor in the Asian and Pacific region', in S. Mukedem ed. *The poor in Asia: productivity raising programmes and strategies*, Asian and Pacific Development Centre, Kuala Lumpur: 1–36.

Murray, M. J. (1980) *The development of capitalism in Colonial Indochina, 1870–1940*, University of California Press, Los Angeles.

Murrey, C. (1986) 'EC-ASEAN trade relations: a Lloyds' perspective', in R. H. Taylor and P. C. I. Ayre eds. *ASEAN-EC economic and political relations*, School of Oriental and African Studies, London: 147–56.

Musimgrafik (1979) *Where monsoons meet: a history of Malaya*, Grassroots Publishers, London.

Myint, H. (1972) *South East Asia's economy-development policies in the 1970s*, Penguin Books, Harmondsworth.

Myrdal, G. (1968) *Asian drama: an inquiry into the poverty of nations*, three volumes, Penguin Books, Harmondsworth, Middlesex.

Nakahara, J. and Witton, R. A. (1971) *Development and conflict in Thailand*, Cornell University Press, Ithaca.

Naval Intelligence Division (1943) *Indo-China*, Geographical Handbook Series B.R. 510.

(1944) *Netherlands East Indies*, Geographical Handbook Series B.R. 518, two volumes.

Nawani, M. A. (1971) 'Punitive colonialism: the Dutch and Indonesian national integration', *Journal of South East Asian Studies*, 2 2: 159–68.

Naya, S. (1987) 'Economic performance and growth factors of the ASEAN countries', in L. G. Martin ed. *The ASEAN success story: social, economic and political dimensions*, East-West Centre, Honolulu: 47–87.

NEDC (National Economic Development Council) (1986) *Medium-term Philippine development plan, 1987–92*, Manila.

Neville, W. (1979) 'Population', in R. Hill ed. *South East Asia: a systematic geography*, Oxford University Press, Kuala Lumpur: 52–76.

Ng Chee Yuen and Wagner, N. (1989) 'Privatization and deregulation in ASEAN: an overview', in Ng Chee Yuen and N. Wagner eds. 'Privatization and deregulation in ASEAN', *ASEAN Economic Bulletin*, 5 3: 209–25.

Ng Chee Yuen (1989) 'Privatization in Singapore: disinvestment with control', in Ng Chee Yuen and N. Wagner eds. 'Privatization and deregulation in ASEAN', *ASEAN Economic Bulletin*, 5 3: 290–318.

Ng Pock Too (1986) 'The future of industrial relations: what role for unions?' Paper presented to the Singapore's Futures Conference, Commonwealth Institute, London, 30–31 January 1986.

Ng R. C. Y. (1979) 'The geographical habitat of historic settlement in mainland South East Asia', in R. B. Smith and W. Watson eds. *Early South East Asia*, Oxford University Press, Kuala Lumpur: 263–72.

Ng Shui Meng (1982) 'Demographic change, marriage and family formation: the case of nineteenth century Nagcailan, the Philippines', Ph.D dissertation, University of Hawaii.

Nidhi, Aeusrivongse (1982) *Bourgeoisie culture and literature of early Bangkok*, Thaikhadi Research Institute, Thammasat University (in Thai).

Nitz, Kiyoko Kurusu (1984) 'Independence without nationalists? The Japanese and Vietnamese nationalism during the Japanese period, 1940–45', *Journal of South East Asian Studies*, 15 1: 108–33.

Oki, Akira. (1984) 'The dynamics of subsistence economy in West Sumatra', in A. Turton and S. Tanabe eds. *History and peasant consciousness in South East Asia*, National Museum of Ethnology, Osaka: 267–92.

(1988) 'The transformation of the South East Asian city: the evolution of Surabaya as a colonial city', *East Asian Cultural Studies*, 27 1–4: 13–48.

O'Malley, W. J. (1979) 'Indonesia in the great depression: a study of Sumatra and Jogjakarta in the 1930s', unpublished Ph.D thesis, University of Cornell 1977, University Microfilms 1979.

Onghokham (1975) 'The residency of Madiun: priyayi and peasant in the nineteenth century', unpublished Ph.D dissertation, Yale University.

Ongkili, J. P. (1977) 'Fillipino opposition to Spanish power: 1868–1898', in Koo Kay Kim ed. *The history of South-East, South and East Asia*, Oxford University Press, Kuala Lumpur: 222–9.

Ooi Jin Bee (1976) *Peninsular Malaysia*, Longman, London.

Osborne, M. (1969) *The French presence in Cochinchina and Cambodia: rule and response*, Cornell University Press, Ithaca.

(1970) *Region of revolt: focus on South East Asia*, Penguin, Harmondsworth.

Owen, N. G. (1978) 'Textile displacement and the status of women', in G. P. Means ed. *The past in South East Asia's present*, Canadian Council for South East Asia Studies, Ottawa: 157–70.

(1987a) 'The paradox of nineteenth-century population growth in South East Asia: evidence from Java and the Philippines', *Journal of South East Asian Studies*, 18 1: 45–57.

ed. (1987b) *Death and disease in South East Asia: explorations in social, medical and demographic history*, Oxford University Press, Singapore.

Pakkasem, P. (1973) *Thailand: North East development planning: a case study in regional planning*, National Economic and Social Development Board, Bangkok.

Palmer, I. (1978) *The Indonesian economy since 1965: a case study of political economy*, Frank Cass, London.

Palmier, L. (1962) *Indonesia and the Dutch*, Oxford University Press, London.

Pang Cheng Lian (1971) *Singapore's People's Action Party*, Oxford University Press, Singapore.

Pangestu, M. and Boediono, S. (1986) 'Indonesia: the structure and causes of manufacturing sector protection', in C. Findlay and R. Garnaut, eds. *The political economy of manufacturing protection: experiences of ASEAN and Australia*, Allen and Unwin, London: 1–47.

Parish, Sir George (1914) 'Great Britain's investments in other lands', *Journal of the Royal Statistical Society*, 71: 456–80.

Parkinson, C. (1960) *British intervention in Malaya, 1867–1877*, University of Malaya Press, Singapore.

Parmer, J. N. (1989) 'Health and health services in British Malaya in the 1920s', *Modern Asian Studies*, 23 1: 49–71.

Parry, J. H. (1963) *The age of reconnaissance: discovery, exploration and settlement, 1450–1650*, Weidenfeld and Nicolson, London.

Pasqual, J. C. (1897) 'Chinese tin-mining in Selangor', *Singapore Journal IV*.

Peagram, N. (1976) 'We don't stay where we are not wanted', *Far Eastern Economic Review*, 92, 14: 2 April 1976: 10–12.

Pelzer, K. J. (1963) 'The agricultural foundation', in R. McVey ed. *Indonesia*, Yale University Press, New York: 118–54.

Penang Gazette, 12 July 1871.

Pendleton, R. L. (1941) 'Laterite and its structural uses in Thailand and Cambodia', *Geographical Review*, 31: 177–202.

(1962) Thailand: aspects of landscape and life, Duell, Sloan and Pearce, New York.

Philippines, National Census and Statistics Office (1971) *Survey of family income and expenditure*, Manila.

(1979) *Integrated survey for households*, Manila.

Philippines, National Economic Development Authority *Philippine Statistical Yearbook*, Manila, various issues.

Pike, D. (1985) 'The security situation in Indo-China', in C. A. Buss ed. *National security interests in the Pacific*, Hoover Institution Press, Stanford, California: 180–96.

Pirani, S. (1987) 'The "struggle" front', in Communist League of Australia, *Vietnam and Trotskyism*: 27–32.

Pires, Tomé (1515) *The Suma Oriental of Tomé Pires*, translated by A. Cortesao, 2 volumes, Hakluyt Society, London, 1944.

Platt, D. C. St M. (1986) *Britain's investment overseas on the eve of the First World War*, Macmillan, Basingstoke.

Polak, J. J. (1943) *The national income of the Netherlands Indies*, Institute of Pacific Relations, New York.

Purcell, V. (1965) *The Chinese in South East Asia*, Oxford University Press, Kuala Lumpur.

Quazi, A. M. A. (1988) 'Planning and development of resource frontier regions in Peninsular Malaysia: achievements and the future', *Third World Planning Review* 10 4: 343–370.

Rada, J. (1982) *The structure and behaviour of the semiconductor industry*, United Nations Centre for Transnational Corporations, Geneva.

Ramachandra, G. P. (1977) 'Vietnamese nationalism', in Khoo Kay Kim ed. *The history of South East, South and East Asia*, Oxford University Press, Kuala Lumpur: 240–59.

Ramsey, J. A. (1976) 'Modernization and centralization in Northern Thailand', *Journal of South East Asian Studies*, 7 1: 16–32.

Ransome, D. (1970) 'The Berkeley Mafia and the Indonesian massacre, in B. Garrett and K. Berkeley eds. *Two, three, many Vietnams*, Cranfield Press, San Francisco: 132–56.

Rao, Y. S. (1984) 'Forest resources of tropical Asia', in Sahabat Alam Malaysia

Environment, development and natural resource crisis in Asia and the Pacific, Penang: 71–85.

Rees, J. (1978) 'On the spatial spread and oligopolistic behaviour of large rubber companies', *Geoforum*, 9: 319–30.

Reid, A. (1980) 'The structure of cities in South East Asia fifteenth to seventeenth centuries', *Journal of South East Asian Studies*, 2 2: 235–50.

(1987) 'Low population growth and its causes in pre-colonial South East Asia', in N. G. Owen ed. *Death and disease in South East Asia*, Oxford University Press, Singapore: 33–47.

(1988) *South East Asia in the age of commerce, 1450–1680*, Yale University Press, New Haven.

Resnick, S. A. (1970) 'The decline of rural industry under export expansion: a comparison among Burma, Philippines and Thailand, 1870–1938', *Journal of Economic History*, 30: 51–73.

Resnick, S. A. (1973) 'The second path to capitalism: a model of international development', *Journal of Contemporary Asia*, 3 2: 133–48.

Rich, E. E. (1967) 'Preface', in E. E. Rich and C. H. Wilson eds. *The economy of expanding Europe in the 16th and 17th centuries*, Cambridge University Press: i–xi.

Richter, H. V. and Edwards, C. T. (1973) 'Recent economic development in Thailand', in R. Ho and E. C. Chapman eds. *Studies of contemporary Thailand*, Australian National University, Canberra: 17–66.

Richter, L. (1986) *The politics of tourism in Asia*, University of Hawaii Press, Honolulu.

Robequain, C. (1944) *The economic development of French Indo-China*, Oxford University Press, London.

Roberts, J. M. (1985) *The triumph of the West*, Guild, London.

Robison, R. (1985) 'Industrial strategies and port development: the Asian case' *Tijdschrift voor Econ en Soc. Geografie*, 76 2: 133–42.

(1986) *Indonesia: the rise of capital*, Asian Studies Association of Australia, Canberra.

(1987) 'After the gold-rush: the politics of economic restructuring in Indonesia in the 1980s', in R. Robison, K. Hewison and R. Higgott eds. *South East Asia in the 1980s: the politics of economic crisis*, Allen and Unwin, London: 16–51.

Robison, R., Higgott, R. and Hewison, K. (1987) 'Crisis in economic strategy in the 1980s: the factors at work', in R. Robison, R. Higgott and K. Hewison eds. *South East Asia in the 1980s: the politics of economic crisis*, Allen and Unwin, Sydney: 1–15.

Rodan, G. (1985) 'Industrialisation and the Singapore state in the context of the New International Division of Labour', in R. Higgott and R. Robison eds. *South East Asia: essays in the political economy of structural change*, Routledge and Kegan Paul, London: 172–94.

(1987) 'The rise and fall of Singapore's "second industrial revolution"', in R. Robison, K. Hewison and R. Higgott eds. *South East Asia in the 1980s*, Allen and Unwin, London: 149–76.

Rostow, W. W. (1960) *Politics and the stages of growth*, Cambridge University Press.

Rowley, A. (1988) 'Waning of the NICs', *Far Eastern Economic Review*, 18 August: 79.

(1989) 'The protection racket', *Far Eastern Economic Review*, 2 March: 93.

Rudner, M. (1975) 'The Malaysian quandary: rural development policy under the First and Second Five Year Plans', in D. Lim ed. *Readings on Malaysian economic development*, Oxford University Press, Kuala Lumpur: 80–8.

Sacks, I. M. (1960) 'Marxism in Vietnam', in F. N. Trager ed. *Marxism in South East Asia*, Stanford University Press: 102–70.

Sadka, E. (1968) *The protected Malay states*, Oxford University Press, Kuala Lumpur.

Sahabat Alam Malaysia (1984) *Environment, development and natural resource crisis in Asia and the Pacific*, Penang.

Saham, J. (1980) *British industrial investment in Malaysia, 1963–71*, Oxford University Press, Kuala Lumpur.

(1980) 'British industrial investment' in L. Hoffman and Tan Siew Lee eds. *Industrial growth, employment and foreign investment in Peninsular Malaysia*, Oxford University Press, Kuala Lumpur.

Salamanca, B. S. (1968) *The Filipino reaction to American rule 1901–1913*, The Shoe String Press.

Salmon, C. (1981) 'The contribution of the Chinese to the development of South East Asia: a new appraisal', *Journal of South East Asian Studies*, 12: 260–275.

Sansom, L. (1970) *The economics of insurgency in the Mekong Delta*, MIT Press, Cambridge, Mass.

Santos, M. (1979) *The shared space: the two circuits of the urban economy in underdeveloped countries*, Methuen, London.

Sar Desai, D. R. (1977) *British trade and expansion in South East Asia*, Allied Publishers, New Delhi.

Sargent, A. J. (1907) *Anglo-Chinese commerce and diplomacy*, Clarenden Press, Oxford.

Sarasin Viraphol (1977) *Tribute and profit: Sino-Siamese trade, 1652–1853*, Harvard University Press, Massachusetts.

Schrieké, B. J. ed. (1929) *The effect of Western influence on native civilization in the Malay Archipelago*, Kolff, Batavia.

Schrieké, B. J. (1955) *Indonesian sociological studies*, part 1, van Hoeve, The Hague.

Schulz, T. W. (1964) *Transforming traditional agriculture*, Yale Press, New Haven.

Scott, A. J. (1987) 'The semi-conductor industry in South East Asia: organisation, location and international division of labour', *Regional Studies*, 21 2: 143–60.

Scott, J. C. (1976) *The moral economy of the peasant*, Yale University Press.

(1984) 'History according to winners and losers', in A. Turton and S. Tanabe eds. *History and peasant consciousness in South East Asia*, Senri Ethnological Studies No. 13, National Museum of Ethnology, Osaka.

Seawood, N. (1988) 'The Daim stewardship', *Far Eastern Economic Review* 141 35: 52–5.

Sen, S. P. (1962) 'Indian textiles in South East Asian trade in the seventeenth century', *Journal of South East Asian History*, 1 2: 92–110.

Shigeru, Ikuta (1988) 'Emergence of cities in maritime South East Asia from the second century BC to the seventeenth century', *East Asian Cultural Studies*, 27 1–4: 1–12.

Shimizu, H. (1988) 'Dutch–Japanese competition in the shipping trade on the Java–Japan route in the inter-war period', *South East Asian Studies*, 26 1: 3–23.

Shoebridge, P. (1988) 'Major commodities of Asia and the Pacific: petroleum', in *The Far East and Australia*, Europa, London: 59–61.

Siddayao, C. M. (1978) *The off-shore petroleum resources of South East Asia: potential conflict situations and related economic considerations*, Oxford University Press, Kuala Lumpur.

Siegal, L. (1979) 'Microelectronics does little for the Third World', *Pacific Research*, 10 2: 15–21.

(1980) 'Delicate bonds: the global semiconductor industry', *Pacific Research*, 11, 1L 1–26.

Sieh, M. L. (1988) 'Malaysian workers in Singapore', *The Singapore Economic Review*, 23: 101–111.

Silcock, T. H. (1967) 'The promotion of industry and the planning process', in *Thailand: social and economic studies in development*, Australian National University Press, Canberra: 258–288.

Singapore, Economic Committee (1986a) *The Singapore Economy: new directions*, Ministry of Trade and Industry, Singapore.

Singapore, Economic Committee (1986b) *A review of Singapore's 1980–84 performance*, Singapore.

Singapore, Economic Committee (1986c) *Causes of the recession in Singapore*, Ministry of Trade and Industry, Singapore.

Singapore, Fiscal and Financial Policy Sub-Committee (1986) *Report to the Economic Committee*, Ministry of Trade and Industry, Singapore.

Singapore, Ministry of Trade and Industry *Economic survey of Singapore*, Singapore National Printers, various issues.

Singh, I. J. B. ed. (1982) *Indians in South East Asia*, Sterling Publishers, New Delhi.

Singhal, D. P. (1960) *The annexation of Upper Burma*, Singapore University Press.

Sithiporn Kridakara (1970) *Some aspects of rice farming in Siam*, Siam Society, Bangkok.

Skinner, G. W. (1958) *Leadership and power in the Chinese community in Thailand*, University of Cornell Press, Ithaca.

Smith, M., McLoughlin, J., Large, P. and Chapman, R. (1985) *Asia's new industrial world*, Methuen, London.

Smith, P. C. and Ng Shui Meng (1982) 'The components of population growth in nineteenth-century South East Asia: village data from the Philippines', *Population Studies*, 36: 253–5.

Smith, R. B. (1978) 'The Japanese in Indo-China and the coup of March 1945', *Journal of South East Asian Studies*, 9 7: 268–301.

Smith, R. B. (1983) *An international history of the Vietnam War, volume I, revolution versus containment, 1955–61*, Macmillan, London.

(1985) *An international history of the Vietnam War, volume II, the struggle for South East Asia 1961–65*, Macmillan, London.

(1988) 'Some contrasts between Burma and Malaya in British policy in South East Asia, 1942–1946', in R. B. Smith and A. J. Stockwell eds. *British policy and the transfer of power in Asia's documentary perspectives*, School of Oriental and African Studies, London: 30–76.

Smyth, H. W. (1898) *Five Years in Siam*, 2 volumes, John Murray, London.

Solheim II, W. G. (1972) 'An earlier agricultural revolution', *Scientific American*, ccvi, 4: 34–41.

Somersaid Moertono (1968) *State and statecraft in old Java*, University of Cornell Press, Ithaca.

Spate, O. H. K. and Trueblood, I. W. (1942) 'Rangoon: a study in urban geography', *Geographical Review*, 32 1: 56–73.

Spencer, J. E. (1973) *Oriental Asia: themes towards a geography* Prentice-Hall, Englewood Cliffs, New Jersey.

Spykeman, N. J. (1942) *America's strategy in world politics, New York: the United States and the balance of power*, Harcourt, Brace and Company, New York.

Sricharatchanya, P. (1983) 'Hospitality can hurt', *Far Eastern Economic Review*, 12 May: 60–1.

Stamp, L. D. (1956) *Asia: a regional and economic geography*, Methuen, London.

Stargardt, J. (1983) *Satingpra I: the environmental and economic archaeology of South Thailand*, Institute of South East Asian Studies, Singapore.

Stauffer, R. B. (1985) 'The Philippine political economy: (dependent) state capitalism in the corporatist mode', in R. Higgott and R. Robison eds. *South East Asia: essays in the political economy of structural change*, Routledge and Kegan Paul, London: 241–65.

Steinberg, D. J. (1981) *Burma's road towards development: growth and ideology under military rule*, Westview, Boulder, Colorado.

Steinberg, D. J. ed. (1987) *In search of South East Asia: a modern history*, University of Hawaii Press, revised edition.

Stenson, M. (1970) *Industrial conflict in Malaya: prelude to the communist revolt*, Oxford University Press, Kuala Lumpur.

Sternstein, L. (1984) 'The growth of population of the world's pre-eminent "primate city": Bangkok at its bicentenary', *Journal of South East Asian Studies*, 15 1: 43–68.

Steward, J. (1977) *Evolution and ecology*, University of Illinois Press.

Stifel, L. D. (1973) 'The growth of the rubber economy of Southern Thailand', *Journal of South East Asian Studies*, 4 1: 105–32.

Stikker, D. and Hirona, R. (1971) 'The impact of foreign private investment', in Asian Development Bank, *Annual Report*, Manila: 370–411.

Strega, J. J. (1983) '"First catch your hare": Anglo-American perspectives on Indochina during the Second World War', *Journal of South East Asian Studies*, 14 1: 63–78.

Stutter, J. O. (1959) *Indonesianisai*, Cornell Southeast Asia Program Data Paper No 36, University of Cornell Press, Ithaca.

Subrahmanyam, S. (1988) 'Commerce and conflict: two views of Melaka in the 1620s', *Journal of South East Asian Studies*, 19 1: 62–79.

Sundaram, J. K. (1987) 'Economic crisis and policy response in Malaysia', in R. Robison, K. Hewison and R. Higgott, *South East Asia in the 1980s: the politics of economic crisis*, Allen and Unwin, London: 113–48.

Suwankiri, T. (1970) 'The structure of protection and import substitution', *Journal of Economics*, 5 1: 53–109.

Takaya, Y. (1987) *Agricultural development of a tropical delta*, University of Hawaii Press, Honolulu.

Tan, E. A. and Holazo, V. (1979) 'Measuring poverty incidence in a segmented market: the Philippine case', *Philippine Economic Journal*, 42, 18 4: 450–92.

Tan, G. (1982) *Trade Liberalization in ASEAN*, Institute of South East Asian Studies, Discussion Papers, Number 32, Sinagpore.

Tan, N. A. (1986) 'The Philippines: the structure and causes of manufacturing sector production', in C. Findlay and R. Garnaut eds. *The political economy of manufacturing protection: experiences of ASEAN and Australia*, Allen and Unwin, London: 48–76.

Tan, T. K. ed. (1967) *Sukarno's guided Indonesia*, Jacaranda, Brisbane.

Tanabe, S. (1978) 'Land reclamation in the Chao Phraya delta', in Y. Ishii ed. *Thailand: a rice growing society*, University Press of Hawaii, Honolulu: 40–83.

Tanabe, S. (1977) 'Historical geography of the canal system in the Chao Phraya delta', from the Ayutthaya period to the Fourth Reign of the Ratanakosin Dynasty', *Journal of the Siam Society*, 65, 2: 23–72.

Tarling, N. (1975) *Imperial Britain in South East Asia*, Oxford University Press, Kuala Lumpur.

(1978a) 'Atonement before absolution: British policy towards Thailand during World War II', *Journal of the Siam Society*, 66 1: 22–65.

(1978b) 'Rice and reconciliation: the Anglo-Thai Peace negotiations of 1945', *Journal of the Siam Society*, 66, 2: 50–112.

(1982a) 'A new and better cunning: British wartime planning for post-war Burma', *Journal of South East Asian Studies*, 13, 1: 9–32.

(1982b) 'An "Empire Gen": British wartime planning for post-war Burma', *Journal of South East Asian Studies*, 13, 2: 310–48.

Tate, D. J. M. (1971) *The making of modern South East Asia, volume one, The European conquest*, Oxford University Press, Kuala Lumpur.

(1979) *The making of modern South East Asia: volume two, The Western impact: economic and social change*, Oxford University Press, Kuala Lumpur.

Tawney, R. H. (1966) *Land and labour in China*, Beacon Press, London.

Taylor, R. H. and Ayre, P. C. I. (1986) *ACEAN-EC economic and political relations*, School of Oriental and African Studies, London.

Tempelman, G. J. (1977) 'Colonization and transmigration in the outer islands of Indonesia', paper presented at the Annual Conference of the Institute of British Geographers, University of Hull, October 1977.

Thailand, Ministry of Agriculture, *Agricultural Statistics*, Bangkok, various issues.

Thaw, Tin Hla (1977) 'The Anglo-Burmese wars: a new look', in Kay Kim Khoo ed. *The history of South East and South and East Asia*, Oxford University Press, Kuala Lumpur: 186–204.

Thoburn, J. T. (1977) *Primary commodity exports and economic development: theory, evidence and a study of Malaysia*, Wiley, London.

Thompson, V. (1947) *Labour problems in South East Asia*, Yale University Press, New Haven.

Thrift, N. (1986a) 'The geography of international disorder', in R. J. Johnston and P. J. Taylor eds. *A world in crisis: geographical perspectives*, Blackwell, London: 12–67.

(1986b) 'The internationalization of producer services and the integration of the Pacific Basin property market', in M. Taylor and N. Thrift eds. *Multinationals and the restructuring of the world economy*, Croom Helm, London: 105–41.

Tjeng, L. T. (1985) 'The power balance as seen from Jakarta: a projection for the 1980s and beyond', in C. A. Buss ed.: *National security interests in the Pacific*, Hoover Institute Press, Stanford, California: 191–206.

Töpfer, B. (1974) 'Zu einigen Grundfragen des Feudalisms. Ein Diskusionsbeitag', in *Feudalismus*, H. Wunder ed. Nymphenburger Verlagshandlung, Munich.

Törnquist, O. (1984) *Dilemmas of Third World Communism: the destruction of the PKI in Indonesia*, Zed, London.

Turley, W. S. (1986) *The second Indochina war*, Westview Press, Boulder, Colorado.

Turnbull, N. (1982) ' "A new and better cunning": British wartime planning for post-war Burma, 1942–43', *Journal of South East Asian Studies*, 13: 33–59.

Unger, L. (1944) 'The Chinese in South East Asia', *Geographical Review*, 34: 196–217.

United Nations (1961) *A proposed industrialisation for the state of Singapore*, Geneva.

Demographic year book, Geneva, various editions.

Year book of international trade statistics, New York, various editions.

United States Embassy, *Indonesia's petroleum sector*, Jakarta, annual.

Valdepenãs, V. B. (1970) *The protection and development of Philippine manufacturing*, Manila University.

Valencia, M. J. (1986) *South East Asian seas: oil under troubled waters*, Oxford University Press, Oxford.

Vandenbosch, A. (1976) 'Indonesia, the Netherlands and the New Guinea issue', *Journal of South East Asian Studies*, 7, 1: 103–18.

van de Koppel, c. (1946) 'Eenige Statistische Gegevans over de Landbouw in

Nederlandsch-Indie', in C. J. J. van Hall and C. van de Koppek eds. *De Landbouw in den Indischen Archipel*, van Hoeve, Gravenhage, four volumes.

van der Eng, P. (1988) 'Marshall aid as a catalyst in the decolonisation of Indonesia, 1947–49', *Journal of South East Asian Studies*, 19, 2: 335–52.

van der Heide, J. H. (1906) 'The economical development of Siam during the last half century', *Journal of the Siam Society*, 3: 74–101.

van Helten, J-J and Jones, G. (1989) 'British business in Malaysia and Singapore since the 1870s', in R. P. T. Davenport-Hines and G. Jones eds. *British business in Asia since 1860*, Cambridge University Press.

van Leur, J. C. (1955) *Indonesian trade and society*, van Hoeve, The Hague.

van Niel, R. (1960) *The emergence of the modern Indonesian elite*, van Hoeve, The Hague.

(1963) 'The course of Indonesian History', in R. McVey ed. *Indonesia*, Yale University Press, New Haven: 272–308.

van der Wijst, C. A. (1983) 'An evaluation of recent Indonesian transmigration programmes', paper presented at the Anglo-Dutch Seminar on migration, held at Soesterberg, 14–16 September.

Varga, E. (1968) *Problems of contemporary capitalism*, Progress Publishers, Moscow.

Värynen, R. (1988) 'East-West rivalry and Regional conflicts in the Third World formations', in B. Heetne ed. *Europe dimensions of peace*, Zed, 1988.

Vasiliev, L. S. and Stuchevski, I. A. (1967) 'Three models for the origin and evolution of pre-capitalist societies', *Soviet Review*, 8: 26–39.

Vatikiotis, M. (1987) 'Resettlement rethink: the government overhauls its transmigration scheme', *Far Eastern Economic Review*, 29 October: 28.

(1989) 'Government yields the lead in Indonesia's development', *Far Eastern Economic Review*: 9 February: 43–9.

Vavilov, N. I. (1926) *Studies on the origin of cultivated plants*, Institute of Applied Botany and Plant Breeding, Leningrad.

von Siam Prinz Diplock (1907) *Die lanwirtschaft in Siam*, Tübingen.

Voon Phin-Keong (1976) *Western rubber planting in South East Asia*, University of Malaya Press, Kuala Lumpur.

(1977) 'American rubber planting enterprise in the Philippines, 1900–1930', *Occasional Papers*, Department of Geography, School of Oriental and African Studies, London.

Wales, H. G. Quarich (1934) *Ancient Siamese government*, Bernard Quarich, London.

Wallerstein, E. (1974) *The modern world system: capitalist agriculture and the origins of the European world economy in the sixteenth century*, Academic Press, London.

Warr, P. G. (1984) 'Export promotion via industrial enclaves: the Philippines' Bataan Export Processing Zone', ANU, Canberra.

(1986) 'Malaysia's industrial enclaves: benefits and costs', in T. G. McGee *et al. Industrialisation and labour force processes: a case study of Peninsular Malaysia*: 179–216.

Warren, C. (1985) 'Class and change in rural South East Asia', in R. Higgott

and R. Robison eds. *South East Asia: essays in the political economy of structural change*, Routledge and Kegan Paul, London: 128–47.
Wertheim, W. F. (1947) 'The Indo-European problem in Indonesia' *Pacific Affairs*, 20: 290–8.
(1959) *Indonesian society in transition: a study of social change*, van Hoeve, The Hague.
West, P. and Alting von Geusau, F. A. M. eds. (1987) *The Pacific Rim and the western world*, Westview, London.
West, T. T. (1982) *Income distribution and determination in West Malaysia*, Oxford University Press, Kuala Lumpur.
Wheatley, P. (1955) 'The Golden Chersonese', *Transactions of the Institute of British Geographers* 21: 61–78.
White, B. (1973) 'Demand for labour and population growth in colonial Java', *Human Ecology* 1 (3): 217–36.
Wilson, Dick (1988) 'European attractions: East Asian manufacturers set up factories in Britain', *Far Eastern Economic Review*, 25 August: 60–1.
Wilson, H. E. (1978) *Social engineering in Singapore*, Singapore University Press.
Wittfogel, K. A. (1957) *Oriental despotism: a comparative study of total power*, Yale University Press, Yale.
Wolf, E. (1969) *Peasant wars of the twentieth century*, Faber, London.
Wolf, E. R. (1982) *Europe and the people without history*, University of California Press, Berkeley.
Wong, J. (1979) *ASEAN economies in perspective: a comparative study of Indonesia, Malaysia, the Philippines, Singapore and Thailand*, Macmillan, London.
Wong Lin Ken (1964) 'Western enterprise and the development of the Malayan tin industry to 1914', in C. D. Cowan ed. *Economic Development of South East Asia*, London: 103–27.
(1965) *The Malayan tin industry to 1914*, University of Arizona Press, Tucson.
(1978) 'Singapore: its growth as an entrepot, 1819–41', *Journal of South East Asian Studies*, 9 1: 50–84.
Woodman, D. (1962) *The making of modern Burma*, Cresset Press, London.
Woon, Tok Kin (1989) 'Privatization in Malaysia restructuring or efficiency disinvestment with control', in Ng Chee Yuen and N. Wagner, 'Privatization and deregulation in ASEAN., *ASEAN Economic Bulletin*, 5, 3.
World Bank (1954) *The economic development of Malaysia*, Washington.
(1959) *A public development programme for Thailand*, John Hopkins Press, Baltimore.
(1963) *Report on the economic aspects of Malaysia*, Government Printers, Kuala Lumpur.
(1976) *The Philippines: priorities and prospects for development*, Washington.
World Development Report, Washington.
(1980a) *Aspects of poverty in the Philippines: a review and assessment*, Washington.

(1980b) *Thailand: towards a development strategy of full participation*, Washington.

(1981) *Indonesia: selected issues of industrial development and trade strategy*, Jakarta.

(1984) *Indonesia: policies and prospects for economic growth and transformation*, Jakarta.

(1985) *Indonesia: policies for growth and employment*, Jakarta.

(1986) *Thailand: Industrial development strategy in Thailand*, Washington.

(1988a) *The Philippine poor: what is to be done?* Washington.

(1988b) *Indonesia: the transmigration program in perspective*, Washington.

Wyatt, D. K. (1982) *Thailand: a short history*, Yale University Press, New Haven.

Yeo Kim Wah (1973) 'The anti-federation movement in Malaya, 1946–8', *Journal of South East Asian Studies*, 4 1: 31–51.

Yuen Choy Leng (1978) 'The Japanese community in Malaya before the Pacific War: its genesis and growth', *Journal of South East Asian Studies*, 9 1: 163–77.

Zimmerman, C. C. (1931) *Siam: first rural survey*, Bangkok Times Press, Bangkok.

(1937) 'Some phases of land utilization in Siam', *Geographical Review*, 27: 373–93.

Index

Abaca 201
Aceh 37
Agency houses 97, 183
Agriculture (*see also* Food and agriculture and Rice)
  contribution to production 24, 25 (table 1.14)
  labour force 25 (table 1.14, 26)
Aid 5
  American to Thailand 175, 176
Angkor 36, 40, 41
Anglo-French rivalry 80–2
Annam 43
Annamite Kingdom 41
Anti-Fascist People's Freedom League (AFPFL) 140, 141
Aquino, Benigno 15
Aquino, Corazon 206
Arab influence 45
Asiatic Mode of Production 47–9
Asian Regional Division of Labour 168, 219–23
Association of South East Asian nations (ASEAN)
  and Kampuchea 10
  as a focus of international capitalism 4, 9
  attitudes towards China 10
  attitudes towards Vietnam 10–11
  defence expenditure 11 (table 1)
  dissent and unity amongst 9–10
  division of labour within 218–22
  founding of 3
  trade within 18, 20, 23 (table 1.13)
  views of USSR 11
Aung San 141
Ayuthia 40, 41
  early European trade with 68
Balance of Payments 31 (figure 1.4)
  in Indonesia 200
  in the Philippines 206
Bananas 151, 201
  multinationals in the Philippines 202–3
Bangkok
  entrepôt trade with China 72–3
  size in 1850 51
  subsidence in 208
Bao Dai 146
Britain 66–9
  agreement with Dutch over spheres of influence 73
  annexation of Burma 76–7
  emergence as the leading industrial power 71
  establishment of control over Malay States 73–4
  growth of interest in South East Asia 66–7, 71–2
  growth of trade with China in the eighteenth century 66–7
  importance of the trade with China 77–3
  occupation of Penang 67
  opening of trade with Thailand 80–1
  rivalry with French over Thailand 80–2
  seizure of Dutch territory 67
  spread of trade into South East Asia after 1800 73
Bumiputeras, promotion of interests 182, 186, 188, 189, 190
Burma
  annexation 76–7
  attraction for traders 76
  conflict with Britain over Arakan 76
  decolonisation 140–1

excluded territories 77
growth and expansion from the 1780s 76
growth of rice exports 91–2
ideology of 3
possibility of re-engagement with international capital 218
resistance to the British 77
Burmese National Army (BNA) 140

Capital exports (*see also* Investment and Foreign investment) 96–7
agency houses' role 97
Capitalist relations of production
degree of establishment in pre-colonial period 55–6, 91
establishment of 96–120
role of the colonial state 97–8
Campa 40, 41
Chenla 40
China, effect of opening to international capital 218, 226
Chinese
capital 112, 113 (figure 4.5), 130, 183
discrimination against 172
estate ownership 96, 101–2
influence of 43, 45
in manufacturing 114
in rubber 103, 104
in teak 96, 107, 108
in tin 95–6, 108
ownership of rice mills 112
role in class formation 129–30
role in the establishment of capitalist production 109–14, 111 (table 4.6), 113 (figure 4.5)
role in trade 45, 96, 97
Class formation
during the colonial period 83, 130
pre-colonial 46–56
Cocoa 183
Coconut oil 105
Coffee 86 (figure 4.1), 87 (figure 4.2), 102, 103
Colonial expenditure
on infrastructure 90, 92, 97, 98
on welfare 82, 132, 133, 134
Colonial legacy 120–38
Colonial revenues 62, 98, 134
Colonial state 97–8
Colonies, self government within 136
Colonisation 61–2, 66, 67 (map 3.4), 69–84 (maps 3.5 and 3.6)
Commodity prices (*see also* individual commodities, Primary production and Terms of Trade) 24 (figure 1.2), 30, 31
Communal organisation 49–50
Community and state 49–52
Copra 105, 201
Culture system 85–7

Debt (*see* Foreign debt)
Debt-led growth in the Philippines 204–5
Decolonisation (*see also* individual countries, Nationalist movements and Nationalism) 138–48
Burma 140–1
colonial attitudes towards 135–7
Indo-China 145–8
Malay States and the Straits Settlements 143–5
NEI 141–3
Philippines 139
role of the Japanese occupation 139, 146
role of the USA 138, 139, 142
role of war time destruction 139
Defence expenditure 11 (table 1.5), 12, 150
Deforestation (*see* Forestry and Timber)
Dependency 84
Dienbienphu 147
Domestic capital (*see also* Chinese, Indigenous, Indian, and Investment)
conflict of interest in Singapore 160
frustration in Thailand 173
in Malaysia 183
promotion of in Indonesia 190, 192
state direction of in Singapore 162
Domestic Industrial Groups
in Singapore 162–3
in Thailand 180–1
Domestic investment (*see* Domestic capital)
Dutch (*see also* Holland) 63–6
at Batavia 65–6
establishment of trade 64, 65 (map 3.3)
fragmentation of the island states 66
nationalist struggle against Spain 64
rivalry with the British 66, 67
spread of territorial control 66, 67 (map 3.4)
Dutch East Indies Company
abolition 66
changing composition of trade 85
establishment and operation 64–5

Eastern Europe
effect of opening to international capital 218, 226

East India Company 67
East Timor, dispute over 9
Economic crisis 30, 33
Economic growth 7 (table 1.3), 33 (table 1.17)
Electronics
  and the ARDL 220, 221 (table 6.10)
  establishment of silicon chip making in Singapore 168
  in Malaysia 187
  in the Philippines 205
  in Singapore 168, 172
England, emergence as the core of the world economy 63
Environmental policies 209
Ethical policy 133, 135
Ethnic division of labour 114
  in Malaysia 181, 182 (table 5.8)
European Community (EC)
  investment from 15, 16 (table 1.7)
  trade with 18, 20–1 (table 1.11)
European World Economy 63–4
Exports (*see* Trade)
Export Oriented Industrialisation (EOI) (*see also* Export Promotion, Export Processing Zones and Industrial Development)
  adoption of 152–3
  conflict with EOI in Thailand 180
  establishment in Singapore 160
  establishment in Thailand 176–7
  in Malaysia 186
  in the Philippines 202, 205
  limited use of domestic raw materials 154
Export Processing Zones (EPZs)
  criticisms of 156
  in Indonesia 196
  in Malaysia 187
  in the Philippines 202
  in Thailand 177
  Singapore economy as 163
Export Promotion
  cost of 156
  ineffectiveness of 156
  similarity of strategies 154–6 (table 5.2)

Federation of Malaysia 144–5
Feudalism 47, 48, 49
Financial srvices (*see also* Service sector)
  growth and promotion of in Singapore 165–6 (table 5.4)
Fishery products 151
  stock depletion 208
Food and agriculture (*see also* Agriculture) 38–42
Food prices (*see also* Rice Premium)
  in the Philippines 204
Foreign capital (*see* Foreign investment)
Foreign debt 32 (table 1.18), 33
  in Thailand 170
  in the Philippines 204–5
Foreign investment 12–17 (tables 1.6, 1.7, 1.8 and figure 1.1)
  and political stability 12–14
  conflict with domestic interests 195, 196
  dependency of development strategies on 207, 222
  disinvestment during 1980s 13
  displacement of Indonesia 190
  incentives (*see also* EOI and Industrial development) 13
  importance of ASEAN in global investment flows 12–23, 13–14
  in Indonesian oil and other primary production 194
  in Malaysia 186
  in Singapore 163, 172
  in Thailand 173, 176, 180
  level in the Philippines 205–6
  outflow of funds 15
  origin of 15 (table 1.7)
  promotion in Indonesia 196
  role in Singapore 13
  sectorial distribution 15 (table 1.8)
Foreign ownership
  in Indonesia 192–3, 194
  in Malaysia 183
  in the Philippines 202
  of the Thai economy 177–8 (table 5.6)
  role in the Singapore economy 13
Forestry (*see also* Teak) 96, 107, 108
  exhaustion of reserves 208
France 77–78
  attempted annexation of Thailand 68
  establishment of Indo-China 77–80
  missionary activity 68–69, 77, 78
  rivalry with British over Thailand 80–2
Free Trade Zones (FTZs) 187
  limited production of value added for the Malaysian economy 187
Funan 36, 40

Geneva Conference 147
Geo-political situation 218

Hemp 104
Ho Chi Minh 146
Holland (*see also* Dutch and NEI)
  emergence as the core of the European World Economy 63

Hong Kong, as a source of investment 15
Housing, in Singapore 161–2, 215
Hydraulic society 47

Import Substituting Industrialisation (ISI) (*see also* Industrial development)
implementation from 1950s 151
in Malaysia 185–6, 186
in Singapore 157–9
in Thailand 172, 173, 17, 186
limitations of 151
maintenance of 152
Incorporation into the world economy
and the rural sector 98–101
summary 217
Independence movements (*see* Decolonisation, Nationalist movements, Nationalism)
Indian immigrants 96, 109, 112, 114
Investment 113 (see figure 4.5), 130
role in the Burmese economy 111, 114
Indian influence 42–3
Indianized states 43
Indigenous capital (*see also* domestic capital, Chinese and Indian) 130
promotion of in Indonesia 190
Indigo 85
Indo-China 77–80
decolonisation 145–8
growth of rice exports 90–2 (figure 4.3)
taxation in 98
Indo-Chinese Communist Part (IPC) 135
Indonesia 190–200
Industrial development (*see also* EOI and ISI) 151–6
colonial period 114–15, 118 (Table 4.7), 119–20
conflict of strategies in Thailand 180
environmental consequences 208–9
in Indonesia 190–1
post-independence commitment to 151–2
promotion, in Thailand 175, 176, 177
role of Chinese capital 115
strategies 152–6
Industrial estates, in Thailand 177
Informal sector 28
Insurgency (*see* Rural insurgency)
Integration into the world economy, degree of 4, 18, 20 (table 1.10)
Singapore 157
Thailand 172, 179
Internationalisation 4
Singapore 157
Thailand 172, 181
Inter-war recession 133
Investment (*see also* Capital Exports and Colonial Expenditure USA, Chinese capital, Indian) 113 (figure 4.5), 151)
growth of in Singapore 168
incentives in Singapore 167
in Indo-China 118 (figure 4.7)
in NEI 101
in plantations 103
Irrigation
failure to develop in Thailand 93–4
in Java 132
pre-colonial 40–4
Islam, spread of 45

Japan
as a source of investment 15, 16 (table 1.7)
conflict with the colonial powers 137
fostering of independence movements 137–8
increasing influence of 12
rivalry with the USA during the 1930s 138
role in US strategy in Western Pacific 10
trade with 18, 20, 21 (table 1.11)

Kampuchea, prospects of re-engagement with international capital 218, 222
Khmer Empire 36, 40

Labour costs (*see also* wage levels) 4, 6 (table 1.2)
and the ARDL 219, 220, 221 (table 6.1)
and the informal sector 28
increase in Singapore 170
in Indonesia 200
in Thailand 181
Labour exports 7
Languages 43 (map 2.3)
Landholding, concentration of 99
Landlessness 99
Labour productivity
in the Philippines 206
Laos, prospects for re-engagement with international capital 218, 222
Lee Kuan Yew 13, 144, 162

Malacca (Melaka) 37
founding of 58
population in sixteenth century 59
Portuguese seizure 59
starved of trade by the Dutch 65
volume of trade in 1510 59

Malacca, Straits of 37, 58
Malay Emergency 143–44
Malay States
ceding of northern states from Thailand 75
Chinese immigration into 74–5
decolonisation 143–5
establishment of British control 73–5, 73 (map 3.6)
tin mining 74–5, 95–6
Malaya, General Labour Union 143
Malaya, Union of 143
Malayan Communist Party (MCP) 143
Marcos, Ferdinand 13, 206
Marx, Karl 47, 49
Mercantilism and the mainland states 68–9
Military involvement in government 12
Mode of Production *see* Asiatic, Capitalist, Feudal and Tributary

Nationalisation 151, 172–3, 190, 200
Nationalism
in Indonesia 190, 191
in Thailand 172
Nationalist movements 139 (see also Decolonisation and Nationalism) 140–1, 142, 143, 145–6
emergence during the 1930s 134–5, 137
National planning
establishment in Thailand 174
Netherlands East Indies (NEI)
decolonisation 141–3
establishment of 66–7 (map 3.4)
plantations in 101
production in 85–7 (table 4.1)
New Economic Policy (NEP) 182, 185, 188, 189, 190
New International Division of Labour (NIDL)
integration of Singapore into 157
integration of South East Asia into 153
Newly Industrialising Countries (NICs)
loss of cost advantage by 220, 225
plans to achieve status, in Malaysia 222, in Thailand 188
plans to achieve status and conflict with intensification of ARDL 220
Oil and natural gas 85, 87 (figure 4.2, 151)
disputes over offshore exploration 9–10
exhaustion of reserves 208
foreign investment in 194
importance to Singapore 164, 165 (figure 5.2)
in Indonesia 192
in Malaysia 183
Pertamina 195
role in development of the Thai Eastern Sea Board 180
role of revenues 195
Oil-palm 85, 87 (figure 4.2), 103, 151, 183
Organisation of production (*see* Production)
Oriental despotism 47

Palembang 37
Partai Komunis Indonesia (PKI) 192
Pathet Lao 147
Penang 37
People's Action Party (PAP) 144, 160, 162, 163, 167, 168, 172
Phibun Songkram 172, opposition to regime of 173, end of regime of 174
Philippines 200–6
acquired by the USA 83–4
decolonisation 139
expansion of the economy under the USA 84, 105–6
growth of export crops under the Spanish 104
nationalist movement 83
occupation by the Spanish 61–2
opening of Manila to international trade 83
religious orders' land-holding 105
resistance to the American occupation 83
Plantations 101–6, 102 (table 4.4), 151
sale of 13
breakup of 151
nationalisation of 151
Political divisions
in the fifteenth century 46 (map 2.4)
colonial legacy of 121
Political stability 13–14
Pollution 208–9
Population 26, 28 (table 1.16), 29 (table 1.17)
growth during colonial period 121–3
Portuguese
eclipse by the Dutch 65
establishment of trade by 59–61
sailing routes 60 (map 3.2)
lack of territorial control 60 (map 3.2), 61
opposition to 62–3
trade with Ayuthia 68
Poverty 209–13 (tables 5.12 and 5.13), 214 (table 5.15)

and land holding in the Philippines 203
during the colonial period 131–4
increased depth and incidence in Indonesia 197–8
Indonesian policies towards 197
ineffectiveness of policies towards in the Philippines 204
Pre-colonial states 36, 37, 39 (map 2.2), 40–1 (map 2.4), 45–6
changing systems of production 53–6
class formation 46–6
extraction of surplus 50–1
living standards 52
organisation of production 46–9
specialist production 51–2
trade (*see also* individual commodities 51–2, 57–8
urban centres 51, 59
Primary production (*see also* individual commodities, Commodity prices and Terms of trade) 4, 7 (table 1.4), 8 (table 1.4b), 18, 86–9 (figures 4.4 and 4.5), 101–6 (table 4.4), 110 (table 4.5), 151
Privatisation
in Malaysia 189
in Singapore 171
Producer services 156–7 (*see also* Financial services)
in Singapore 166–7
Production, in the pre-colonial period
changing systems of 53–6
organisation of 46–9
Protection (*see also* Tariffs) 153–4 (table 5.1)
levels in the Philippines 205
Malaysian tariffs 186
pressure to reduce in the Philippines 205
Thai tariffs 175, 176
Public expenditure (*see* State expenditure)

Racial division of labour (*see* Ethnic division of labour)
Raffles, Sir Stanford 236
Rebellions 134
Recession of the 1970s and 1980s 30–1, 33–4
impact on Singapore 168–9 (figure 5.3)
Regional planning (*see also* Spatial imbalance)
in Thailand 174–5
Regional structures (*see also* Asian Regional Divisions of Labour) 1, 3
Resource depletion 207–8
Restructuring of the world economy, impact of 33–4, 217–18
Rice cultivation (*see also* Rice production)
and population density 41
and the environment 38 (figure 2.1)
and pre-colonial cultures 39–42
and water control 40–2
intensification of 41
yields, Burma 100
yields, Thailand 94 (figure 4.4)
Rice exports
from Java to the spice islands 52
growth of mainland after 1850 89–95 (figure 4.3)
in seventeenth century 51
price fall 1929–31 100
Rice Premium (Thailand) 150
Rice Production, post-independent expansion
in Indonesia 197
in Malaysia 184–5
in the Philippines 203–4
Role of the State (*see* Colonial state, State expenditure, State involvement in the economy, State involvement in production)
Rubber 85, 87 (figure 4.2), 103–4, 183, 192
Rural development 149
and anti-insurgency in Thailand 175
in Malaysia 185, 188
in the Philippines 203
Rural insurgency 149
in Thailand 174
in the Philippines 203
Rural sector, role of 150
Rural unrest during 1930s 134

Sarit Thanatrat, Marshal 174
Service sector (*see also* Producer services)
investment in 15, 17 (table 1.8), 26–7
Singapore 157–72
development since 1950s 157–72
dependence on foreign investment 13
rise as an entrepôt 73
role in the NIDL 218
role in the ARDL 219, 220
Skilled labour (*see also* Training)
shortage in Thailand 181
shortage in Malaysia 188
Small-holding 103, 106
South East Asia
division between pro-and anti-Western states 3

South East Asia (*cont.*)
level of development 21–2
recognition as a region 1–2
resource endowment 4, 7 (table 1.4a)
significance to the world economy 4, 6, 7, 9
structure of the economies 24, 25 (table 1.14)
Spanish
occupation of the Philippines 61–2
rivalry with the Portuguese 61
trade routes 60 (map 3.2), 62
Spatial imbalance (*see also* Regional planning)
effectiveness of policies 209, 212–3 (table 5.15)
in Indonesia 192, 194–5, 196, 198, 199 (table 5.11)
industrial policy in Malaysia 188
policies towards 207–8
Spatial pattern of economic activity (*see also* Spatial imbalance and Regional planning)
and foreign investment 15
during colonial period 123–7
Spice 52, 57, 58
areas of production 58 (map 3.1)
attraction for European traders 59
fifteenth-century trade to Europe 58
Javanese control of production
Portuguese trade in 61
State expenditure
in Indonesia 197 (table 5.10)
in Malaysia 183, 184 (table 5.9), 185, 189
in Singapore 167
level of 33, 171 (table 5.5)
State involvement in the economy
Indonesia 190, 191, 195
the Philippines 204
State involvement in production
in Indonesia 192
in Malaysia 182, 186
in Singapore 162–3
in Thailand 173
Structural adjustment 33–4
in Indonesia 200
in the Philippines 205, 206
in Thailand 33, 179
Subsistence sector 98–9
as an adjust to rubber cultivation 103
Suez Canal, impact of 87
Suharto 13, 193
Sukarno 192, 193
Srivijaya 36, 37
State participation in the economy 223
Malaysia 183
Sugar 87 (figure 4.2
Indo-China 104
Java 85, 86
Philippines 8, 104, 105
Singapore 167
Thailand pre-1855 88 (table 4.1)
Sunda Straits
as a routeway 65 (map 3.3), 66
closure of 9
Swettenham, Sir Frank 94

Taiwan, investment from 15
Tariffs (*see also* Protection)
as a source of revenue in Thailand 172
Malaysia 186
opposition to in Malaysia 185
Tea 104
Teak (*see also* Timber and Forestry) 88 (table 4.1), 89 (table 4.2), 96, 107, 108
Technical transfer 220–1
Technology, introduction of European 68, 69
in tin mining and smelting 108–10
Tenancy
absence of 191
growth 99, 100 (table 4.3)
Terms of Trade 30 (figure 1.3), 31
Thailand 80–2, 172–81
as a new mainland core 222
Bowring Treaty 81
Burney Treaty 80
changing composition of trade 87–9 (tables 4.1 and 4.2)
changing land-use 88
concessions to the West 70 (map 3.5), 81–2
goal of NIC status 222
growth of rice exports 89–90, 92–5
influx of Western enterprise after 1855 88
irrigation, failure to develop in early 1900s 93–5
reduction of British influence 138, 145
rice yields 93–4 (figure 4.4)
Timber (*see* Forestry and Teak)
Tin 74–5, 102 (figure 4.2), 95–6, 108–9, 192
Chinese capital in 74–5, 95–6, 108
Western capital in 95–6, 108–9
Tobacco 88 (table 4.1), 104, 183, 201
Tomé Pires 58–9
Tonlé Sap 36, 40
Tourism 7

Trade (*see also* Terms of Trade, Commodities and individual commodities) 18–21 (tables 1.9, 1.10, 1.11), 22 (table 1.12), 23 (table 1.13)
  changing composition in the nineteenth century 85, 87–9, (tables 4.1 and 4.2)
  composition of in Singapore 164, 165 (figure 5.2)
  composition of in Thailand 181 (table 5.7)
  growth of in the Asian NICs 164 (Figure 5.1)
  in the 1930s 116–17 (figure 4.6)
  intra-regional 18, 23 (table 1.12)
  with Japan 18, 20 (table 1.11)
  with the EC 18, 20 (table 1.11)
  with USA 18, 20 (table 1.11)
Trade Unions
  control of in Singapore 160–1, 167
  repression in the Philippines 202
Training of the labour force
  in Singapore 162, 167
Transmigration in Indonesia
  since independence 198, 199 (table 5.11)
  under the Dutch 124, 132
Tributary mode of production 49
Trotskyists in Vietnam 135

United Malay Nationalist Organisation (UMNO) 143
USA
  and the Philippines 82–4, 200–1
  and Vietnam war 147–8
  as a source of foreign investment 15 (table 1.7)
  conflict with the colonial powers 137, 138
  establishment of Pacific Island bases 82
  expansion of trade in Asia during the nineteenth century 82
  investment 104, 105
  military aid to Thailand 175, 176
  military presence in Thailand 175
  military strategy 9, 10, 218
  rivalry with Japan during the 1930s 138
  role in decolonisation 138, 147
  seizure of the Philippines 82–4
  trade and investment in Thailand during 1930s 138
  trade with 18, 20–1 (table 1.11)
Urban centres
  colonial 126–6
  management of 215
  pre-colonial 37, 39, 41, 51, 59
  redevelopment of 215
Urban growth 213
  environmental consequences 208–9
  ineffectiveness of controls over 215–16
Urbanisation 26, 27 (table 1.15)
Urban primacy 26, 27 (table 1.15)
USSR
  ASEAN and US views of 10–11
  treaty with Vietnam 10
  use of Vietnamese ports 11, 12

van der Heide, J. H. 93
Vasco da Gama 59
Vietminh (Viet Nam Doc Lap Dong Minh) 146, 147
Vietnam
  invasion of Kampuchea 10–11
  naval bases 11
  prospects for re-engagement with international capital 218, 222
  relations with ASEAN 10
  relations with USSR 9
  strategic importance of 10
VOC (Vereenigde Oostindische Compagnie) *see* Dutch East Indies Company

Wage levels (*see also* (Labour costs)
  rise in Singapore 1979–81 167
Wittfogel, Karl 4